Alpheus Baker Hervey

Sea Mosses

A Collector's Guide and an Introduction to the Study of Marine Algae

Alpheus Baker Hervey

Sea Mosses
A Collector's Guide and an Introduction to the Study of Marine Algae

ISBN/EAN: 9783337417598

Printed in Europe, USA, Canada, Australia, Japan

Cover: Foto ©berggeist007 / pixelio.de

More available books at **www.hansebooks.com**

SEA MOSSES

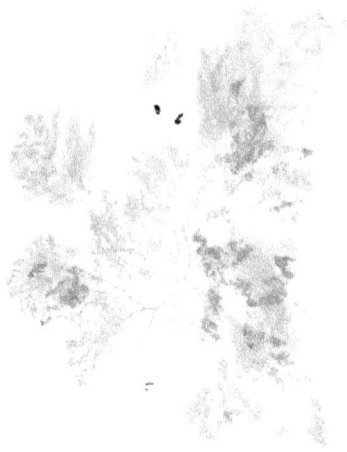

HAMNION VERSICOLOR,
HAMNION HETEROMO...

COLLECTOR'S GUIDE

AND

AN INTRODUCTION TO THE STUDY

OF

MARINE ALGÆ.

BY

A. B. HERVEY. A.M.

ILLUSTRATED WITH TWENTY FULL-PAGE ENGRAVINGS IN COLOR,
FROM PHOTOGRAPHS OF ACTUAL SPECIMENS.

BOSTON:

BRADLEE WHIDDEN.

1893.

TO

RICHARD HALSTED WARD, M.D.,

PROFESSOR OF BOTANY

IN THE RENSSELAER POLYTECHNIC INSTITUTE,

TROY, NEW YORK;

IN THE NAME OF A LONG AND TRUE FRIENDSHIP;

AND IN APPRECIATIVE RECOGNITION

OF A NATURALIST,

DISTINGUISHED ALIKE FOR CLEARLY APPREHENDING,

AND SKILFULLY IMPARTING

THE TREASURES OF A SCIENTIFIC SCHOLARSHIP,

SINGULARLY WIDE AND EXACT;

THIS BOOK

IS AFFECTIONATELY INSCRIBED

BY

THE AUTHOR.

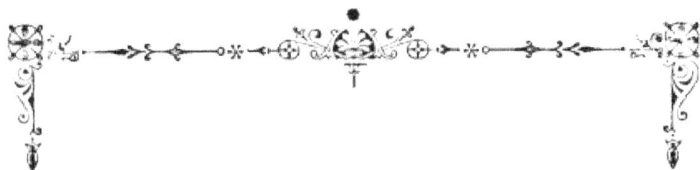

PREFACE.

———:o:———

I AVAIL myself of the last opportunity which I shall have for a word with my readers to add a point or two to what will be found on *p. 4, et seq.,* of the "Introduction," concerning the method of this book. I have attempted to make a book which should be a real and helpful guide to those, who, though not expert botanists, and not having, or using, any aids to a good pair of eyes other than a simple pocket magnifier, desire to begin the collection and study of marine plants. I have been obliged, therefore, to resort to many devices for making the novitiate see for the first time in these plants what is so obvious to the practiced eye of the experienced collector.

Among these is the particular thing which I wish to direct attention to here, viz.: the disarrangement of the species in the genera. It will be observed that while the genera have been arranged in their proper natural order, the species are often grouped in the text quite otherwise. The reason is, I have taken those species, in genera which contain several for treatment first, which, on account of their commonness, or peculiar habitat or appearance, could be most easily and certainly identified. From these I have proceeded step by step to the more difficult plants.

Then again I have often found it convenient to group certain species together for the advantage of comparison in the description which do not always naturally belong together. ' You will therefore understand that, while the orders and genera follow their natural grouping, in the text, the species in the genera cannot be depended upon to do so, in most cases.

I must add a single remark further on this general subject. While the several sub-classes, the Green,

Olive Colored, and Red Algæ, are grouped in the ascending natural order, in the text, the orders and genera in each of them are arranged and treated in exactly the opposite order, the first being the most highly, and the last the most simply, organized genus in each sub-class.

I must take this occasion to express my large indebtedness to several fellow students of Algæ, for help in making ready the material for this book. To the published notes, the private correspondence, and personal assistance of Dr. Wm. G. Farlow of Harvard University, I am under very many obligations. I can only regret, for my readers' and my book's sake, that I could not avail myself of all the new knowledge contained in his Manual of New England Algæ, which is now long overdue from the Government Press.

Prof. Daniel C. Eaton, of Yale College, has been ever kind, obliging, and painstaking, allowing me to draw without stint upon his ample store of knowledge, and his well-furnished herbarium.

Mr. Frank S. Collins, of Malden, whose acquaintance with the marine flora of Massachusetts Bay is both extensive and accurate; Mrs. Maria H. Bray, of Magnolia, and Mrs. Abbie L. Davis, of Gloucester, who have long been known as careful students and industrious collectors about the rocky and fertile shores of Cape Ann; and Miss M. A. Booth, of Long Meadow, who has spent several summers of profitable collecting on the east end of Long Island, have each kindly made out for me lists of the plants which they have collected in their several localities, together with notes of their special habitat, season of growth, and frequency of appearance.

Dr. C. L. Anderson, of Santa Cruz, Cal., Dr. N. L. Dimmick and Mrs. R. F. Bingham, of Santa Barbara, and Mr. Daniel Cleveland, of San Diego, all well-known collectors and Algologists, have very obligingly done the same thing for the plants of their several localities on the Pacific coast. In addition to that, they have sent me many valuable typical specimens from the rich and extremely interesting flora of that region.

Nor can I forget the generous assistance which for years past I have received from that veteran collector in New York waters, Mr. A. R. Young, of

Brooklyn. I have the memory of many delightful excursions about the shores of New York Bay in company with him, who knows so well when and where all the finer and rarer plants are to be had. I am permitted to quote him all too seldom in these pages because the light has been shut out — let us hope only temporarily — from those eyes which were ever so keen to detect, and so appreciative in recognition of, the rare beauties of these humble, but exquisite forms.

If this book shall be of any service to any in opening the way to a knowledge of this department of Botany, or shall contribute anything to the pleasures of summer life by the Sea-side, no small part of the merit must be accorded to our enterprising publisher, Mr. S. E. Cassino, at whose urgent solicitation the work was undertaken, and who has spared no pains or expense to make it as valuable and acceptable as possible.

The plates for this volume are engraved from photographs of specimens in my herbarium. In outline and color, therefore, they represent real plants.

It is with no small degree of solicitude that I send

forth this little book upon its mission. The best wish I can have for it is that it may impart to its readers a tithe of the pleasure its preparation has given to its author. I may, perhaps, be allowed to hope, that it shall communicate some interesting knowledge to many inquirers, and awaken in many appreciative minds an intelligent admiration for this part of Nature's wondrous handiwork.

A. B. HERVEY.

TAUNTON, MASSACHUSETTS,
  May, 1st, 1881.

LIST OF PLATES.

—:o:—

TABLE OF CONTENTS.

———·o◊o·———

CHAPTER III.

OLIVE COLORED ALGÆ.

CHAPTER IV.

RED ALGÆ.

I heard, or seemed to hear, the chiding Sea
Say, Pilgrim, why so late and slow to come?
Am I not always here, thy summer home?
Is not my voice thy music, morn and eve?
My breath thy healthful climate in the heats,
My touch thy antidote, my bay thy bath?
                    Behold the Sea,
The opaline, the plentiful and strong,
Yet beautiful as is the rose in June;
Creating a sweet climate by my breath,
Washing out harms and griefs from memory
And, in my mathematic ebb and flow,
Giving a hint of that which changes not.
I with my hammer, pounding evermore
The rocky coast, smite Andes into dust,
Strewing my bed, and, in another age,
Rebuild a continent of better men.
Then I unbar the doors: my paths lead out
The exodus of nations; I disperse
Men to all shores that front the hoary main.

Emerson.

CHAPTER I.

—:o:—

INTRODUCTION.

On the surface, foam and roar.
Restless heave and passionate dash;
Shingle rattle along the shore,
Gathering boom and thundering crash.

                  *   *   *

Under the surface, loveliest forms,
Feathery fronds with crimson curl,
Treasures too deep for the raid of storms,
Delicate coral and hidden pearl.

CHAPTER I.

INTRODUCTION.

> There is a pleasure in the pathless woods,
> There is a rapture on the lonely shore,
> There is society where none intrudes,
> By the deep sea, and music in its roar.
> I love not man the less, but nature more,
> From these our inteviews, in which I steal
> From all I may be, or have been before,
> To mingle with the universe, and feel
> What I can ne'er express, yet cannot all conceal.
>
> *Byron.*

WHO does not love the sea! For every mood of the mind, with some one of its thousand voices it speaks some answering tone. Those who dwell within the sound of its surf, or those who habitually seek its presence for inspiration of soul, or for rest and health of body, learn to love it for its own sake and for its sweet and comforting companionship. I know what those feel who are content to sit for hours

beside the sounding sea, and watch the incoming and
outgoing tides, as

> "The nightmared ocean murmurs and yearns,
> Welters and swashes, and tosses and turns,
> And the dreary black seaweed lolls and wags;"

or listen listless to the beating of the sleepless waves,
as they go tumbling among the rocks,

> "With sobs in the rifts where the coarse kelp shifts,
> Falling and lifting, tossing and drifting,
> And under all a deep, dull roar,
> Dying and swelling for evermore;"

or send their thoughts wandering around the world,
cruising on every shore, with that white sail yonder
which just now slid down behind the edge of the sky.

Somehow, one cannot look upon the wide blue sea,
and listen to its rythmic beating, without feeling
that in some true sense he is looking into Nature's
soul, and hearing her great heart beat. For true it
is, the mighty voice of Old Ocean plays a low melodious
accompaniment to all the deepest thoughts that stir in
the human heart, and makes the soul feel its eternal
kinship with all the great forms and forces of the
universe.

But, there is another pleasure which "this great
and wide sea" can give us, besides that which she
offers to our fancy and our dreams. It is the con-
templation and study of the exquisitely beautiful flora
which she nurtures in her ample waters. When you

know the sea and its flowers, you will know that she
has almost a mother's love and tenderness for them.
It may seem to you a dumb, rude, bungling sort of
affection, perhaps, for you will notice that she often
leaves some delicate and charming flowers, far up on
the hot sand or stones of the beach, all careless if
they live or die. But you will also see that she is sure
to come back to them again by and by. But, in the
sea, where they live and grow, they have her constant
offices of care and nurture. These most fragile fronded
plants, whose silky branches are as fine as the thinnest
cobweb, are handled and tended so gently, that not a
fibre is broken or a cell misplaced in the midst of
pounding waves, which, with a single blow would crush
an iron ship to atoms. The boisterous sea is their
home, and though it may seem rough and rude to
us, it is never ungentle to them.

If you come to know these plants, the beauty, delicacy, and grace of them, and their names, habits, and
history, I am sure the sea will have an added charm
for you. From every shore you visit you will carry
away your hands full of them. And these garlands, in
after years, will not only minister to your love of the
beautiful, but they will also recall the blessed hours
spent by the sea, and repeat in your heart again the
joy of its mighty presence.

In this little book I shall attempt to make you acquainted with what I have, these many years, found to be as interesting as they are beautiful. I undertake the work *con amore.* I remember how much I needed some convenient and competent guide when first I wanted to enter this field of knowledge and delight, and asked in vain for it. I have many friends who often go down by the breezy margin of Old Ocean. With this book I want to make them acquinted with some delightful friends of mine who will be there before them. I have spent many hours of rare pleasure in collecting, mounting, and studying, these simpler forms of Nature's handiwork. I greatly desire to share this delight with the multitude of intelligent people who spend weeks and months by the sea side yearly, and the not less intelligent multitude who make their homes within sound of its waves.

The work is written for beginners only, and not for advanced students and specialists in this department of Cryptogamic botany. I am ambitious for my book that it may be just a "Porter" to stand at the gate of this wondrous garden of the sea, and open for those who come and knock. There was no such book to do this in my day, so I had to "climb up some other way." There were indeed the three ponderous quartos of Harvey, and two or three little manuals of English

Algæ, to be found in the American market. But neither served the needs, at once, of a beginner, and of a sea side rambler upon American shores. I said just now, " for those who come and knock." The " Porter " opens the door only to such in any garden of delight, or palace of good. There must be interest enough to lead one to ask admittance. If you want to go in and see what is growing in this strange world under the sea, you have only to come and knock, and heed what the " Porter " says to you at the gate, and you may go in, and wander far and wide amid the beauties of this charming flora.

To begin with, then, I must assume that you are willing to put a little earnest work into this study. What you achieve with some cost, you will enjoy with more zest. But I shall attempt so to present the matter as to call for the least possible labor in attaining the best results. The descriptions of the plants, will, as far as possible, be confined to those points of appearance etc. which can be seen with the unaided eye, or at least with the help of a simple pocket lens. Especial attention will be given to pointing out the particular kind of place where each plant naturally grows, and the season of the year when it may be found most abundantly, so that you will be able to search intelligently for it, and be all the

more likely to know it when you see it for the first
time.   In making descriptions of the plants, I shall
make use of technical terms only when common
terms cannot be found to answer; or when, without
the technical words, I should have to make circumlocu-
tions which would be burdensome both to you and
to me. The few words of this kind which I shall be
obliged to use, and which are not defined in the
dictionary, will be found in a Glossary at the end of
this volume.

I am aware that there is a popular prejudice
against the use of any other than the common names
for plants and animals. People think it is an affecta-
tion of learning, a very silly pedantry, for these
naturalists to go about and speak of the birds and
flowers and ferns, and call them by such outlandish
"jaw - breaking" names, as they do. But I must
bespeak your favor, to put away this prejudice, at least
in respect to the "Sea Mosses."   If you study these
plants at all scientifically, you will be obliged to learn
their scientific names, and, for the best of all reasons,
because almost all of them have no other. A few like
the   "Dulse," *Rhodymenia palmata;* "Rockweed,"
Fucus nodosus and *F. vesiculosus;* "Irish Moss,"
Chondrus crispus; and "Devil's Aprons" or "Kelp"
Laminaria; have common or popular names. But

the people who have lived by the sea, have, as a
general thing, cared very little for the "Sea
Weeds," and have deigned to give names to but a
few of them. So it has been left to the botanist to
christen them from his Greek and Latin vocabulary.
For each plant he has provided two names, a "sur-
name," and a "given name." The former answers
to the name of the genus, and is the family name;
and the latter is the individual name, or the name
of the species. But he writes it with the generic or
family name first, and the "given name" last. In
his usage it is "Smith John," not "John Smith," as
in common parlance. Thus *Rhodymenia palmata*
and *R. corallina*, may be considered sisters, the first
being the family name, and the last two the "given
names" by which they are known in the family circle.
Do not be discouraged on account of these hard looking names. They are no harder to remember, or to
pronounce, than the names of your personal friends,
Mrs. Eliza Watson Thompson or Mr. George Washington Jones. When from affectionate interest and acquaintance, you are able to number these beautiful
creations of Nature among your friends, you will find
it perhaps easier to recall their names, than those of
your more fashionable acquaintances. For you will
find that these names mean something as a personal

description, which is more than can be said of most
human patronymics. The names of plants are mostly
terms descriptive of some notable fact in their appearance, habit, structure, place of growth, or fruiting.
The significance of the names will, as far as possible,
be indicated as we come to them.

Before passing from this point, I must not forget
to say, that you may be intelligently interested in
these charming plants ; be an admirer of their brilliant
and varied . colors, their graceful outlines, and their
slender and delicate forms ; may, perhaps, be an enthusiastic collector of them, and more deeply in love
with them than many " marble hearted " botanists are,
and yet, never care anything at all about a scientific
knowledge of them, or give them a single hour's scientific study. Scores of people have for years gathered
these " flowers of the sea," and arranged them on
cards, and mounted them in books and albums, who
never knew them other than as " Sea Mosses,"
and never cared to. You may do the same if you
choose. In that case you will find this introductory
chapter all the guide you will need. If you have not
time or inclination to study them, do not neglect them
on that account. To the taste that appreciates the
beautiful in form or color, they are an endless source
of pleasure, and a sure means of cultivation. The

plants of the sea greatly surpass all others in the
perfection with which they retain their original beauty
when dried and preserved in the herbarium. Indeed,
some of them are more beautiful so, if possible, than
when seen in their native element. Their artistic value
will not be impaired by any lack of scientific knowledge
on your part. And yet I must assure you that a more
particular acquaintance with them will abundantly
repay all your labor by giving you a more intelligent
interest in them. And it will make you a better collector, even for the mere beauty's sake, to know the
habits, homes, and seasons of these beautiful creations.

GEOGRAPHICAL DISTRIBUTION.

You will find it an important help, in many cases,
to pay attention to the geographical distribution of
the species, so as not to look for what you cannot
find in given localities, and to search only for what
may reasonably be expected to grow there.

Our eastern coast is distinguished by two quite
well marked floras. That long reach of land which
projects itself so far into the sea, known as Cape Cod,
marks the division between the two. It is probable
that in former times, more than now even, that has
prevented the waters of the great arctic and equatorial
currents from mingling, and so has maintained a

marked difference in temperature, in the two regions.
At all events the floras of the two regions have important differences, whatever the cause. I do not mean
by this that no considerable number of species
extend over the whole region, north and south of
Cape Cod. But I mean that a considerable number,
enough to make a distinct feature of the flora, do
not extend either way beyond that barrier. To state
it broadly, we may say that the plants growing north
of Cape Cod are essentially arctic, and agree pretty
well with the species found on the extreme northern
coasts of Europe, and in Spitsbergen and Nova Zembla.
In a small collection of some twenty species received
from these polar islands, I find all but one or two of
them such as I have collected at Marblehead. The
individual plants, too, have a striking resemblance to
those growing along our northern shores. The northern flora is distinguished by an abundance of plants of
the species *Euthora cristata*, *Ptilota plumosa*, Var.
serrata. *Ceramium Deslongchampsii*, *Gigartina mam-
illosa*, *Halosaccion ramentaceum*, *Fucus furcatus*,
Agarum Turneri, *Laminaria longicruris*, *Alaria
esculenta*, etc.

The flora south of Cape Cod is that of the warmer
or temperate seas, and is distinguished by the presence
of such forms as the "Gulf weed," *Sargassum vulgare*,

Dasya elegans, the several species of the *Chondriopsis,* the *Grinnellia Americana, Rhabdonia tenera, Hypnea musciformis, Champia parvula, Lomentaria Baileyana, Spyridia filamentosa, Collithamnion Baileyi* and many others. I suppose, perhaps, that from one quarter to one-third of the species of each region do not extend into the other, or, if they do at all, then as rarities. I will note the geographical range of each species as I describe it. There seem to be no such differences in the flora of different parts of California. It is likely that nearly all the plants that could be found at San Francisco or Santa Cruz, could also be found at San Diego and Santa Barbara, a few rarities only excepted. It will be observed that this book undertakes to give an account only of the marine flora of California on the west coast, and of New York and New England on the east; though, it may be added, that this will make it practically applicable to all the coast north of the Carolinas on the one side, and to Vancouver Island on the other. I may also add that I have included only common plants, such as the beginner would be certain to meet with in his sea side excursions; and I believe I have included nearly all of these on our eastern shores. I cannot say as much for the California flora. I have selected for special mention only some sixty or seventy species peculiar

to that region, which is much richer in species than
our own. But I have taken those plants which I
judged to be the most common and characteristic,
and most widely distributed, and such as I knew to
be most strikingly beautiful or interesting. In respect
to particular places, there are many of them on our
eastern coast where the flora is rich and fine, and
where thousands of people are in the habit of going
every year. Nothing could be more favorable as
places for finding and collecting splendid "Sea Mosses"
in great numbers and many varieties than such
localities as Mount Desert, the Maine and New Hamp-
shire beaches, Isles of Shoals, Cape Ann from Annis-
squam clear around to Magnolia, Marblehead, Nahant,
Nantasket, Newport, Martha's Vineyard, and Wood's
Holl, Orient Point, and the shore at Coney Island, and
southward as far as Fort Hamilton.

CLASSIFICATION.

Algæ are classified by botanists on the basis of
their method of reproduction. In a popular work of
this kind I have not thought it desirable to enter into
the details of this matter, because these organs can
be studied only by the aid of a microscope ; and,
as I have said, I am writing for those who do not use
that instrument, and I hope to be able to so describe

the plants that most of them may be identified without its aid.

Suffice it to say that the whole class naturally divides itself into three main groups, characterized in a general way by their color, viz.: Red, Olive Green, and Bright Green. These three groups correspond very nearly to their more exact classification on the basis named above. The lowest and simplest in their organization, are the bright or grass green Algæ, for example, the *Ulva*; next the olive green, the " Rockweed " and " Kelp "; the highest, the red Algæ. I shall take up each of these groups separately, and describe the several genera and species, in their natural order, following the arrangement adopted by Dr. Farlow, from Prof. Thuret, in his list of North American Algæ.

TIMES AND PLACES FOR COLLECTING.

Most collecting on our Atlantic coast, will be done during the summer and early autumn months. But I must remind those of you who live by the sea, or have it accessible at all times, that many things of the greatest interest and beauty will be missed if you do not go to the shore early. Our finest *Callithamnion, C. Americanum* can be had in its rarest beauty early in March and even in February. The finest varieties of our *Rhodomela subfusca* are only to be found in the early spring months. This is true of many other plants. You will be surprised, also, to see what quantities of things you can find as late as November and December. Indeed, if you are to know these plants thoroughly, you must collect them at all seasons of the year. Then you will know when they come, and when they go, and when they are in their greatest perfection. Those living and collecting on the Pacific coast are not fenced away by an icy wall, as we are on our shores during two or three months of our hard, inclement winters. So they can collect the year around. Dr. Anderson assures me that most of the plants growing there may be found at all seasons, though of course most of them are more beautiful and of more luxuriant growth during the summer than during the winter months. In general, there are three principal places for collecting "Sea Mosses" by the shore.

First, from the mass of material which the sea throws up upon the beaches, and leaves behind it when the tide goes out. This will be your main resource for getting the plants that grow in deep water. By many causes they will be loosened from their holdings in the depths, and will then float up to the surface and margin of the sea, and will be cast on shore. By carefully turning over these masses, which will be

found along almost every sandy or pebbly beach, you
will be able to get plants which could otherwise be
found only by dredging in the deep water. And by
careful search, too, among this material, you will find
all the deep water forms.

Second, upon the rocks and in the tide pools
when the tide is out. You can collect living plants in
their native homes here only. Of course no Algæ
grow upon the sandy beaches. You must, therefore,
seek all such as grow between the tide marks,
upon rocky shores. Put on a pair of stout rubber
boots, and go two or three hours before low tide
and search in every place, following the tide down
to its farthest retreat. Many of the best things
are found close down by low water mark, and some
a little below that. These latter can be got best
by taking advantage of the extreme low run of tides
which comes about "new" and "full moon." The advantage of going before low tide, and following the
retreating waters down, is that you are not so apt
to get a drenching by the unexpected advance of a
great wave, as when the tide is coming in. For, if
you are close by the water's edge when the tide is
rising, busily intent upon getting your floral treasures,
you will very likely find yourself suddenly soaked
with brine, for

"The breaking waves dash high
On a stern and rock-bound coast."

In hunting through the tidal region for plants, hunt everywhere, and collect everything found growing, and when collected, like Captain Cuttle, "make a note of it." If you cannot remember without, carry a small memorandum book and enter in it the habitat of each particular kind as you collect it. The tide pools, that is, the little basins in the rocks out of which the water is never emptied, are the places where the choicest collecting may be had. And the nearer they are to the low tide limits, the more likely they will be to have abundance of vegetable life in them. But do not fail to look, also, under the overhanging curtain of "Rockweed" which shadows the perpendicular sides of the cliffs and great boulders. You will often find some beautiful plants there, as for instance, the *Ptilota elegans*, the *Cladophora rupestris* and other smaller "mosses."

Third, by standing on some low projecting reef, by the side of which the tide currents rush in and out, you will see many of the more delicate deep water forms, all spread out beautifully, and displayed in all their native grace, carried past, back and forth in the water. Many of these, like the *Poly-siphoniæ*, are seldom thrown on shore in good condition, or if they are, do not long remain so. This therefore is by far the best place to take many of these plants. To do this you must be provided with some simple instrument for reaching down into the water, and seize them as they go floating by. I have found nothing more convenient for this than a wire skimmer, which can be got at any house-furnishing tin shop, tied with a stout string to a light strong stick five or six feet long. The water passes through the meshes of this with little resistance, but the Alga, with its delicate branches thrown out widely in every direction, is very readily caught by it. It will also serve to a limited extent as an implement for detaching plants from their holdings, which grow in deep tide pools, or in the sea, not too far below low water mark. For the rest of your

COLLECTING APPARATUS

you may have as little or as much as is convenient. A simple basket, or box, with a few newspapers in it, to wrap up and keep somewhat separate the different sorts of your collectings, will do very well. If it is convenient, have a case made with a half dozen or less wide-mouthed bottles set in it, each provided with a cork. The case should also have a compartment for storing coarse plants, newspapers, paper bags, or whatever you may use for keeping

different species, or the plants from different localities, separate. Then, as your plants are collected,
they may be roughly sorted, and put in different
bottles. But two or three bottles should be reserved
for the most delicate and fragile forms. And as there
are several of them which rapidly perish on being
exposed to the air, the bottles should be kept partly
full of sea water. The more delicate *Polysiphonias*,
the *Calithamnions, Dasyas*, and some others will need
this protection. I have found a quart fruit jar very
handy. I get the kind that I can fasten a string
around the neck, so as to carry it suspended in
one hand, which leaves the other always free to
gather in the plants with. A jar whose cover goes on
and off with the least possible trouble, is the one to
be selected. The only disadvantage in using a
receptacle of this sort for your collection, is that
in climbing over the wet and mossy rocks, your feet
may chance to slip and you get a tumble; then in
your efforts to save yourself, you will forget all about
your fragile glass jar, and will smash it into a thousand pieces upon the hard stones, and perhaps lose
your whole collection. But two or three of these jars,
carefully packed in a basket, so as not to be easily
broken, would perhaps furnish as handy a collecting
apparatus as you could extemporize at the sea shore.

MOUNTING AND PRESERVING.

For "floating out" your "Sea Mosses," as it is called,
you should provide yourself a few simple tools and re-
quisites. You should have a pair of pliers; a pair of
scissors; a stick like a common cedar "pen stalk," with
a needle driven into the end of it, or, in lack of that,
any stick sharpened carefully; two or three large
white dishes, like "wash bowls;" botanist's "drying
paper;" or common blotting paper; pieces of cotton
cloth, old cotton is the best; and the necessary
cards or paper for mounting the plants on.

You will use the pliers in handling your plants
in the water. The scissors you will need for trimming
off the superfluous branches of plants which are too
bushy to look well when spread upon the paper,
and to cut away parasites. The needle should be
driven point first, a considerable distance into the
stick, so as to make it firm, and allow you to use
the blunt end of it in arranging the finer details of
your plant on the paper. For drying paper, of course
you can use common newspaper, by putting many
thicknesses together; and a great many, no doubt,
will do that. But sheets of blotting paper will be
found much more satisfactory; twenty-five of them
cut into quarters would probably be all you would

use, and those you could easily take with you in your trunk. What will be found cheaper and still more serviceable, if you are going to mount a large number of plants at once, is a quantity of botanist's "drying paper." It can be had of the "Naturalist Agency," 32 Hawley Street, Boston, Mass., for, I believe, $1.25 per 100 sheets, probably also of other sellers of naturalists' supplies in all the large cities, on both sides of the Continent. It is a coarse, spongy, brown felt paper, cut into sheets, 12 x 18 inches, and has a fine capacity for absorbing moisture. For convenience, the cotton cloths should be made the same size as the drying paper used. Some collectors, who do not care to mount a great number of specimens at once, but want to have them very smooth and fine when dry, use no drying paper at all, but in the place of it, have thin smooth pieces of deal, got out a foot or so square and one-quarter or one-third of an inch thick; upon these they spread one or more layers of cotton and lay the plant on them and put as many more over it; the cotton absorbs the moisture, and the boards keep the pressure even and the papers and plants straight and smooth throughout. For "mounting paper" each one must use his own taste. Many prefer cards cut of uniform size : they can be had at almost any

paper store, or job printing office, made'to order. Four
and a half by six and a half inches, is a neat and
convenient size. But if you want to mount several
hundred or several thousand specimens in the course
of a season, so as to have some to give to all your
friends, and to make up a number of books or
albums to sell at Church or Charity fairs, then perhaps the expense will be an item worth considering.
In that case you will find it cheaper to buy a few
quires of good 26 or 28 lb. demy paper, unruled, of
course. This paper is in unfolded sheets, 16 x 21
inches, and will cut into convenient sizes for mounting any plants ordinarily collected. By halving it
you have sheets 8 x 21, or $10\frac{1}{2}$ x 16 inches. By
quartering, the sheets are 8 x $10\frac{1}{2}$ inches; halving
these you get an octavo sheet $5\frac{1}{4}$ x 8 inches,
which is quite large enough for the great ma-
jority of plants. One half of this will give a sheet
4 x $5\frac{1}{4}$ inches, which will be the size most used; while
the smallest plants look best on the half of these
sheets, $2\frac{1}{2}$ x 4 inches.

With your large white dishes filled near to the
brim with sea-water, or, if you are away from the
ocean, with water made artificially salt, take a few
of your plants from the collecting case and put
them in one of the dishes. Here, handling them

with your pliers, shake them out and clean them
of any adhering sand or shells, trim away parasites
and superfluous branches, and generally make them
ready for " floating out." Thence transfer them, one
at a time, as you "float them," to the other dish.
Then take your card, or your paper, selecting a
piece large enough to give the plant ample room,
and leave a margin of white all around and having
dipped it in the water, put it quite under the
floating plant, holding the paper with your left hand
and managing the plant with the right. Now float
the plant out over the paper, and draw the root or
base of it up near to the end of the paper next
your hand, so that you can hold it down on the
paper with the thumb of your left hand, the rest of
that hand being under the paper in the water.
Now slowly lift the paper up to the surface and
draw it out of the water, in such a way that the
water will flow off from it in two or three directions.
This will spread the plant out somewhat evenly over
the paper. But in many cases you will need to
arrange the branches in their most natural and graceful position and also take care that they do not get
massed upon each other, and make unsightly heaps,
while other places are left bare. They should be
carefully arranged so as to make the most beautiful

picture possible. In some fine and delicate plants, too much care cannot be bestowed in having the remote branchlets all naturally disposed and spread out. This final work of arranging details you will do with your needle while you hold the paper very near to the surface of the water with your left hand, so near, indeed, that there will be just water enough and no more, above it, to float the delicate parts which you are manipulating. Oftentimes it will be found convenient, after the paper with the plant on it has been removed from the water, to re-immerse a part of it at a time, and re-arrange the several parts separately. But all this can easily be done, more easily than I can tell how to do it. A very little practice will give you the "knack" perfectly. And, indeed, these plants are by no means refractory, or hard to manage. They will do anything you can reasonably want them to, while you humor them by keeping them in their native element. In fact, you will commonly need to do no more with them than to just help them do what they are altogether willing and disposed to do themselves. For if you will let them take on your paper the form and outline, which they have by nature in the water, there will be nothing left to desire, for their color, form, and movement, all combine there to make them the loveliest

and most graceful things that grow. When you have
put the last finishing touches upon the "floating"
process, and your "Sea Moss" is adjusted upon your
paper so as to be "a thing of beauty, and a joy
forever;" then you want to lay the paper upon some
inclined surface, any smooth board will do, to drain
away the superfluous water. Thence it is to be trans-
ferred, in a few moments, to the press for drying.

This is made in the following manner. Laying
down one of the above described sheets of blotting
paper, botanist's " drying paper," or boards of muslin-
covered deal, you lay your paper with the plant on
it upon this, the plant up. Cover the board or drying
paper all over with "floated" specimens in the same
way. Over all, and lying directly upon the plants,
spread your piece of muslin. Upon this, put another
sheet of the paper, or board, and upon this again,
a layer of plants, then a piece of the muslin, more
paper, plants, muslin, and so on till you have disposed
of all of your collection, or so much of it as you care
to mount. Upon the last layer of plants put a final
sheet of paper, and over all a stout board as large
as the drying paper. Upon this lay some heavy
weights — stones will be as handy as anything at the
sea-side. I should put on, I think, about fifty pounds
of them, if I were using botanist's drying paper,

which has a good deal of "give" in it. With the use of boards unless there are a good many thicknesses of muslin, it would not do to weight it so heavily, or some of the plants would be crushed beyond recognition. I use the drying paper, and always have two boards, one for the bottom, and one for the top of my press. Then, when I "have made the pile complete," I can put it aside in some convenient corner out of the way, and set the stones to work, bearing down on it, a business for which they seem to have some conspicuous and weighty gifts.

Some botanists recommend that the drying papers be changed in the course of five or six hours, and the cloths and papers again in twenty-four hours. This will, perhaps, be best, if one has plenty of time. But my practice has always been to let them lie twenty-four hours, and then give them a change of both cloths and papers, being careful in removing the cloths, so as not to lift the plants from the mounting paper.

The second time in the press they should be subject to a harder pressure, seventy-five or one hundred pounds of stone being not too much. In twenty-four hours more most of them will be quite dry, and ready to be put into your herbarium, album, or whatever you use for the final disposition of them.

Those that are not perfectly dry should be put back in the press with dry papers and cloths for another day's stay.

When the plant is perfectly dry, and removed from the press, you should, before putting it away and forgetting these facts, write on the back of the paper the exact date and place of collecting.

People often ask me what I use to make the plants stick so firmly to the paper, supposing, evidently that it is necessary to have some kind of gum or mucilage for that purpose. I have to answer that I have for most of them to use nothing whatever; that there is sufficient gelatinous matter in the body of the plant to make it perfectly adhere to the paper without other aid. And the reason for putting the muslin over the plants in the process of pressing and drying, is that they may not stick to the drying paper which is laid above them, the muslin not adhering to the plants at all, except in some few cases.

But a considerable number of the "Sea Mosses do not adhere to paper well. They either have not gelatinous matter enough in them, or will not give it out to glue their bodies to the paper. Various devices are resorted to in these cases. Sometimes the plant, after being dried in the press in the usual way, is simply strapped down with slips of gummed paper

Sometimes they are fastened down with some kind of adhesive substance, after being dried, gum tragacanth being the best for this. Others take them and float them out a second time in skimmed milk, and after wiping off the milk from the paper and plants, except directly under the plants, put them in the press to dry again, when, it is said, they stay. I have never tried this method. A friend of mine, who is famous for the artistic way in which she always "lays out" her "Sea Mosses," tells me that for these forms which lack what the Phrenologist might call "Adhesiveness," she prepares from the "Irish Moss," *Chondrus crispus*, a semi-fluid paste, into which she dips them before putting them on paper, and then carefully removes all of it from the paper and plant, except what is between the two, and then puts them in the press. By this means, they are made to stick, "like the paper upon the wall."

In preparing the coarser "Rockweed" and "Kelp" for the herbarium, another method will have to be pursued. These will almost all turn very dark, or quite black, in the process of drying. I am accustomed to treat them according to the following method: Taking them home, I spread them out in some shaded place and let them lie for a few hours, perhaps twenty-four, perhaps less or more,

until most of the water in them has evaporated,
but not till they have become hard, stiff and brittle.
Then I put them between sheets of drying paper
and lay them in the press, and keep them there
until the process of drying is complete. A little
practice will be the only way by which you will
learn how to tell if they have been dried long
enough in the open air. If you find them inclined
to mould while kept in the press, you may be sure
they are not dry enough; throw them away and get
some new ones.

It is sometimes desirable to keep the treasures
we have gathered from the sea unmounted, that we
may carry them away to await a more convenient
season for floating them out, or that we may send
them to some friend or correspondent on the other
side of the continent or beyond the seas. It is,
therefore, fortunate that all but the more delicate
and perishable of these plants may be dried rough;
rolled up, and kept any length of time; transported
round the world; and then, when put in water
again, will come out in half an hour, as fresh and
bright and supple and graceful as they were when
taken from their briny home. The friend just
now referred to assures me that even the *Callithamnia*, *Dasyæ*, and the most delicate *Polysiphoniæ*, and

such like plants, may be so treated, by first shaking
the water out of them and then thoroughly mingling
them with dry sea sand, and drying them rough in the
usual way. She says the sand will adhere to the
most delicate fibres and ramuli of the plant in such
a way as to keep them separate and prevent their
getting glued together. Then when they are afterwards soaked out, the sand will be disengaged and
the plant left as good as ever it was. Perhaps I
ought to suggest that "soaking out" should always
be done with salt water, unless you know you
have only those plants that fresh water will not hurt.
When I have had specimens of the "Rockweed" or
"Kelp" sent me "rough dried," I have found it
best to prepare them for mounting, not by immers-
ing them in water, and so getting a great quantity
of moisture into them, which would have to be ex-
pelled afterwards with no little trouble, but by wrapping them about with wet towels; from these they
would imbibe enough dampness to be manageable,
but not enough to make them troublesome.

Before taking leave of this part of my subject, I
must permit myself to add a word in regard to a
point which botanists commonly think too little about,
viz: the display of taste in the mounting of their
plants. To the mere botanist a plant is a *specimen*

of a given genus and species, interesting wholly ror
that fact. If it is a full grown typical form with
fruit, all the better. Now all are not botanists.
Most of those who will read these pages will have
an interest in these plants to which the scientific
interest will be secondary. I want to say then to
them : look for the best things, get the whole
plant when you can, but get and preserve the
most perfect and beautiful plants. It is the rule
with the botanist to put but one species on each
paper or card ; I certainly advise disregarding this rule,
unless you are mounting for scientific purposes altogether
or chiefly. With the numberless shades of red which
one group of "Sea Mosses" will give you, with the
various kinds of green which the other two will
present, you will have opportunity to display all the
taste and skill you are master of. For in combining
several different colors and forms on the same
paper, you may often produce the most brilliant
results. A little practice will soon make you able to
handle two or three plants at the same time in
" floating them out," almost as readily as you manage one. Then again, you will soon find it possible
with some of the more slender plants to work out
interesting and beautiful " designs " in the same way.
Initial letters, even monograms, may not be beyond

your reach with a little care and practice. Let the
"Sea Mosses" contribute to the cultivation of every
faculty, and all possible means of pleasure for you.

For preserving your treasures after they are neatly
mounted, pressed and dried, you have two courses
open to you. You can take care of them as the
botanist does, by arranging them systematically in a
herbarium, with covers of stout Manilla paper folded
10½ x 16½ inches for each genus, and the species
separated by white sheets or thinner covers; or you
can provide yourself with blank books, made for the
purpose, having the leaves cut to fit the sizes of
paper or card which you mount your plants on, so
as to slip the corners of the cards into the cuts.
It is well in that case to provide a book with
leaves large enough to hold two or four cards each.
By following the directions here given, I cannot
doubt you will soon become a successful collector,
and an expert in mounting and preserving "Sea
Mosses."

METHODS OF STUDY.

Having now the book as you go to the sea shore,
the question you are most likely to ask is: "How
shall I use it, so as to make it a true and helpful
guide in learning about these plants?" I will try to
tell you in a few words. Most of the descriptions

are written from herbarium specimens, and describe them as they appear spread out on paper. And yet where there are characteristic points to be seen when the plant is found growing in its native element, they are mentioned. You will therefore find it particularly serviceable in identifying mounted specimens. And knowing these, you will have little trouble in recognizing them living. But the important question is, how shall you bring the book and the plant together, so as to make the one guide the learner to the other. First of all by paying careful attention to what the book says, for in every instance it puts the emphasis of its description upon the distinguishing mark of the species. In the next place, use your eyes in looking at the plant, and use your powers of mental observation. Do not be of those who "having eyes see not." Now there are, as I conceive, two ways of bringing the book and plant together. The first is by taking a plant and hunting up its description and name in the book. You have two ways for doing this : first, see if the plant in question is figured in any of the plates ; if so, its name is there and it will be easy to find the description. If you do not find it figured, see if you do not find some plant figured which is near enough like the one you are studying to be a brother or cousin to it. If you do, that will give you the name of the genus. Go there, and among the species you will find the plant in question. Suppose, for example, that you have a frond of the *Ptilota elegans* under observation, you will not find that in the plates; but you will find a beautiful copy of a *Ptilota plumosa* var. *serrata*, which you will see much resembles your plant, but is not it. This will lead you to the right genus, and then you will soon have the thing settled.

Again, you will find "keys" at the head of all the great divisions of the book, which if carefully used, will lead you easily to the genus you are in search of, and once there you will readily find the species sought. Suppose, for example, you find a mass of curled and kinky wool-like, green "Sea Moss," floating on the tide or entangled with Algæ on the rocks; looking at it carefully till you observe that it is a simple un-branched thread of green, you turn to the "key" for Green Algæ; the frond is not membranaceous, so you will not find it in the first group. It is filiform, or thread-like, therefore you will find it under one or the other of the sub-division of this group. It is un-branched, so you are sure to find it in the first division, for there you read, "Frond unbranched, sometimes attached straight and single, sometimes float-ing, kinked and matted like wool," which is an account

of the plant you are making inquiries about, and you
find that these plants are in the genus *Chætomorpha*.
Turning now to that, you will find an account of
the plant, such that you will not doubt you have
before you *C. tortuosa.*

A second way of making the book and the plant
meet is to select a few common plants that the book says
may be found anywhere, and carefully noting the
description, and especially its habitat, with the best
image you can form of it in your mind, go to the
places where it ought to grow and there search for it
till you find it.   For example, you will read in the
book that the *Polysiphonia fastigiata* grows upon the
ends of *Fucus nodosus* like little brown or black balls
as big as a walnut.   Now go down and find some o
this *Fucus* and search till you find some with its
parasite on it.   You will read that *Ptilota elegans* just
now referred to, grows common on the perpendicular
sides of cliffs and large rocks, under the curtain of
the overhanging " Rockweed." Go there and hunt till
you find it.   You are told that many plants of the
species *Cystoclonium purpurascens* have little curling
tendril-like branches which twine around other plants;
go down to the shore and turn over the mass
which the retreating tide has left, till you find
some specimens of it, and you will not have to search

long. In this way you may find a great many of the common forms and easily identify them "by the book."

In making your beginning in these studies, take the easiest first; those that are commonest and have easily distinguished marks. From the more easy proceed step by step to the more difficult. Do not spend unnecessary care and labor in trying to make out difficult cases. Put them aside for the present. When you have had more practice it will be easier for you.

Again, you may presume a little on the good nature and kindness of botanists, and especially of Algologists, and send your difficult plants to them to name for you. I have often done such service for people. I thus try to repay the kindness and patience with which my footsteps were guided, when I first set out in this path, by many far more distinguished botanists than I ever expect to be. I have not a little indebtedness of this kind still unliquidated, as I trust some of my readers will take the liberty of finding out.

Still another way to get help, is to get some Algologist to spare you out of his duplicates, by exchange or purchase, some of the forms which you are inquiring about, and thus have something authentic for comparison. You would have very little difficulty then in fixing the place and name of your own plant.

CLUBS AND CLASSES.

Supplementary to the subject presented in the last section, a few words on the formation of Clubs and Classes for the collection, mounting, and study, of "Sea Mosses," may be said. The many advantages of associated over solitary action is everywhere recognized. Everybody knows that in any undertaking where half a dozen people can be engaged together, more interest, enthusiasm, pleasure, and profit, can be derived than where one works all alone. So I want to recommend that when you go to the sea shore with your friends, or go among strangers and make acquaintances and friends at hotels, boarding-houses, or "camps," anywhere indeed, where two or three, or half a dozen, intelligent persons are collected, you set about organizing a "Sea Moss Club." It will not take much talk or enthusiasm on your part to convince some of them at least, that collecting and mounting these "things of beauty" will be a very pleasant and engaging way of spending the leisure hours of a summer sea-side vacation. When it is practicable, each one should be armed with a copy of this book, as the best "Collector's Guide."

You will need no formal organization perhaps, or if
you want to have a name for your extemporized society,
call it after some eminent Botanist. If one of your
number has had experience, or is more wise than the
rest in such things, let him be appointed your leader
or director, and if you care to keep a record of your
doings, of your tramps, adventures, successes, and
failures, your collectings, and your progress, appoint a
"ready writer" for your secretary. Such a record
might sometime be of real value to scientific botanists
in making notes of the flora of the region, and in
finding the habitat of uncommon species. It certainly
would in after years serve to recall many pleasant
memories. For collecting expeditions along the shore,
or to neighboring islands, go all together, or divide
off when it would be best, so as to send parties of
two each, to different localities, thus reaching as
many points as possible. Let each collect for all, that
is, collect enough specimens of each kind so as to be
able to supply all with duplicates. The study of
new or unknown plants, both mounted and unmounted, will be vastly more interesting and profitable, if it is carried on in company with the others.
The saying is, "two heads are better than one, if
one is a sheep's head." So, six pair of eyes and
six thinking minds are surely more than six times as

good as one, in searching the books, and identifying the plants.

I venture to predict, that you will find the doings of the "Sea Moss Club" an extremely pleasant diversion, both socially and intellectually. You will find as a result, that every member will be awakened to a stirring, thrifty, new interest in Nature's things, and has acquired at once a keen appetite for the charms of her more rare and delicate handiwork, and a new faculty for seeing and observing her wondrous ways.

> "Nature hath tones of magic deep, and colors iris bright,
> And murmurs full of earnest truth, and visions of delight;
> 'Tis said, 'The heart that trusts in her, was never yet beguiled,'
> But meek and lowly thou must be, and docile as a child.
> Then study her with reverence high, and she will give the key,
> So shalt thou learn to comprehend the 'secret of the sea.'"

And I shall venture also to believe that, when you

> "Fold your tents like the Arabs,
> And as silently steal away"

from the sound of the surf, and the sight of the sea to take up your toils again in the hub-bub and confusion of this work-a-day world, you will be very sure to keep up the pleasant memories of the "Club," and perhaps also its form, by correspondence, and further study and exchange of plants. And, perhaps, you will hear of other Clubs, formed and working at other points of the coast, and you will enter into correspondence and exchange with them also.

HISTORY.

It would be an interesting branch of the subject if I had the necessary space at my command, to give an adequate historical sketch of the cultivation of this branch of botanical science in America. It would be especially so if I could allow myself to give even a brief account of the most distinguished workers in this field. But I cannot. The enumeration of a few names, dates, and incidents is all I can expect to find room for at this time.

Of course I am not in possession of data by which I can ever tell how many scores or hundreds of people every year employ their leisure hours by the sea-side, in collecting, mounting, and arranging these plants. We know of a few of them who have given their collections to botanists to write about.

The first person who seems ever to have interested himself in American Algæ, was Mr. Archibald Menzies, who singularly enough made his collections on the Pacific Coast. The *Phyllospora* from that coast which bears his name, was described, from plants which he brought from there by the celebrated Dawson Turner, in the early part of this century. He accompanied Vancouver in his expedition to North Western America in 1792-3, and with him sailed around the world.

Harvey speaks of him, as he knew him late in life, as one of the best preserved specimens of a green old age that he ever knew, still enthusiastic in his studies; and with his plants before him, recalling with great vividness the stirring and often adventurous scenes which were associated with their collection. Many of them more than half a century gone. Harvey writes: "It was his enthusiasm which first possessed me with a desire to explore the American shores, a desire which has followed me through life."

In 1825, Beechy made his exploring expedition into the North Pacific and brought home many plants, an account of which was published in 1833. In July, 1840, a Russian exploring expedition touched the California coast, and carried away several interesting plants, some of which were described and figured by Ruprecht, in St. Petersburgh, in 1852. Subsequently Dr. Coulter collected in Monterey Bay.

The first collector of California Algæ, whose collections fell into the hands of botanists, subsequently to the time of the great emigration to that land in '49, was Mr. A. D. Frye, of New York city. His collections were made about 1850. They attracted some attention in New York as well as in San Francisco. The plants in this collection are the ones

chiefly used by Harvey in making his account of
the Pacific Algæ in the "Nereis." Since that time,
and especially during the last ten years, many industrious botanists have been at work on that rich and
beautiful flora. I need not here mention the names
of this distinguished company, for several of the best
known of them get frequent mention in the pages of
this book. These and others appear often in the
botanical publications by other hands.

Previously to 1850, the knowledge of the marine
botany of our eastern coast was in a very imperfect
and chaotic state. There were a few collectors in
Boston and vicinity. How much any of them, with
the exception of Dr. Gray, knew about the natural
history or the systematic arrangement of the plants
does not appear. They included among others such
men as the late Mr. Geo. B. Emerson and Dr. Silas
Durkee. Mr. Stephen T. Olney, of Providence, who
did no inconsiderable work in illustrating the botany
of Rhode Island, collected a large number of Algæ,
which are now in the Olney Herbarium of Brown
University.

A few enthusiastic and capable collectors about
New York city had been at work for some time,
inspired and guided by that able and devoted naturalist,
Prof. J. W. Bailey, of the West Point Military Academy,

whom Dr. Harvey calls "the earliest American worker
in the field of Algology." He sent the first specimens
of our American Algæ to Dr. Harvey. Though
Prof. Bailey lived a considerable distance from the
sea, he was mainly instrumental in awakening an
interest in these plants among those who were better
situated for collecting them than he. They were accustomed to send their plants to him, and when he could
not resolve them after patient study, he sent them abroad
to be determined by the more advanced Algologists
of Europe; and so, gradually, there came to be a little
scientific knowledge about these things diffused among
American collectors. There was a little knot of enthusiastic Algologists in New York city and Brooklyn.
Among them, Hooper, Lounsbury, Pike, Congdon,
Walter and Averill, with whom Bailey was in constant
correspondence, and evidently sometimes went collecting.

In a letter, which I have, written by him to Mr.
Hooper, he refers to that company in a pleasant way
as the "Algerines," and invites them all to come up
to West Point, and look over his collections; "then,"
he says, "I believe you will carry the war into
Barbary with new zeal. It will be no less pleasure,"
he adds, "to show my microscope, &c., to several
friends at the same time than to one alone." In

those days, before 1850— though how much before I cannot say, as the letter has no date—a microscope, in this country at least, was a curiosity of no small moment. Of that company I believe only Captain Pike remains.

A complete set of the published and manuscript notes of Prof. Bailey's patient and accurate scientific observations, together with his scientific correspondence, his large collection of Algæ, and no less than 3,000 mounted and catalogued microscopical objects, are in the possession of the Boston Natural History Society, and are accessible to all students of science.

It was mainly through the influence of Prof. Bailey, that Dr. Wm. H. Harvey, Prof. of Botany in Trinity College, Dublin, and the most learned and distinguished British Algologist, came to this country, to study and publish our plants. Arrangements were made for the publication of the Memoir, and Dr. Harvey came here about 1850, and remained in the country several months visiting important points from Halifax to Key West, and collecting largely, also availing himself of the collections of others. From the material thus gathered, he published through the Smithsonian Institution, the largest work ever yet issued on American Algæ—the "Nereis Boreali-Americana."

The first part containing the olive colored sea

weed, was published in January, 1852; the second
part on the red sea weed, about a year later; and
the third on the green Algæ, not till 1857, after
Dr. Harvey's return from Australia. They are in
quarto form, contain 50 colored plates, and can be
bought for about $25.

Since those days a new generation has come up.
But in the meanwhile, for a space of twenty years,
scarcely anything was published on American Algæ.
At the present time there are a few enthusiastic collectors, and a still smaller number of devoted students
of Marine Algæ scattered up and down our extensive seaboard. The names of several of them will be
found making frequent appearance in these pages.
Only two of our more distinguished living botanists
have given special attention to this subject: Dr. Wm.
G. Farlow, of Harvard University; and Prof. Daniel C.
Eaton, of Yale College; the former of whom brings
to his work the advantage of several years' critical
study of these plants under some of the most celebrated Agologists of Europe — the lamented Thuret,
and the learned Agardh, and others. Dr. Farlow's
publications consist of several annotated lists of Algæ,
including new species, issued in the proceedings of the
Academy of Arts and Sciences, and in the reports of
the U. S. Fish Commissioners. A much more elaborate

work from his pen will shortly be published under the auspices of the Fish Commission, if indeed it shall not come to my readers before they see this.

I cannot conclude this introductory chapter, without saying that if this book shall be the means of awakening any interest in these creations, among the sojourners by the sea-side, I should be sorry if it should fail to carry the mind beyond the creature to the Creator.

To me, the best story which any flower of land or sea can tell, is the story it whispers to my heart, not only of the skill and wisdom which fashioned it, but also of the beneficient and sleepless care which has kept and preserved it, has ministered to its humble wants, and will not let it perish without His notice.

> " Not a flower
> But shows some touch in freckle streak or stain,
> Of his unrivaled pencil."

> " The Lord of all, Himself through all diffused,
> Sustains, and is the life of all that lives,
> Nature is but a name for an effect,
> Whose cause is God; He feeds the sacred fire,
> By which the mighty process is maintained;
> He sleeps not,— is not weary; in whose designs
> No flaw deforms, no difficulty thwarts,
> And whose benificence no change exhausts."

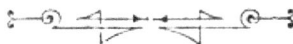

CHAPTER II.

BRIGHT GREEN ALGÆ.

KEY TO THE GENERA.

BRIGHT GREEN ALGÆ.

I. FROND MEMBRANOUS.

    1. Color *Green*.

(*a.*) Frond, wide, long and thin, the largest green Algæ.

                    Ulva.

(*b.*) Frond, narrow, sometimes inflated, always tubular.

                    Enteromorpha.

    2. Color, *Brown or Purple*.

Frond, thin, translucent, sheeny, satin-like.

                    Porphyra.

II. FROND FILIFORM.

    1. Frond *Unbranched*.

Sometimes attached, straight and single, sometimes floating, kinked and matted like wool.

                    Chætomorpha.

    2. Frond *Branched*.

(*a.*) Stem and (straight) branches each a single cell, not jointed.

                    Bryopsis.

(*b.*) Stem and branches jointed, that is, composed of short single cells attached end to end.

                    Cladophora.

Wasser, du Mutter des Lebens. In dunkler Tiefe der Meere,
  Preisen die Wesen all', Fische und jeglich Gewürm
Deine gebärende Kraft; von ihr auch zeugen die Ströme,
  Zeugt noch der Tropfen vom Teich voll microskopisch Gethier,
Und sich nähren wollen sie alle! Siehe, und ihnen
  Wächst auf krystallenem Grund tausendfaltiger Tang,
Gleich Arabeskengewirre gigantische Blätter und Bänder,
  Fluthende Gärten voll Pracht in dem doch lichtlosen Reich.
Ueberall grünt's auch in See'n und Strömen von zarteren
        Pflänzchen,
  Zittert doch selbst noch im Bach zartestes Algengewirr,
Grünende Strähnen glitzernde schlüpfrige Klumpen,
  Deren Wundergehalt sich nur dem Forscher entdeckt;
Staunen erfasst die Seele vor all dem Geheimniss des Lebens,
  Welches das kleinste Gebild selbst noch im Tropfen enthüllt.

CHAPTER II.

DESCRIPTION OF GENERA AND SPECIES.

Sub-Class.— *CHLOROSPORÆ.*
Order.—*SIPHONEÆ.*
Genus.—*BRYOPSIS* Lam.*

THE American genera of this order are all inhabit-
ants of the warmer seas, except the *Bryopsis*,
and that is represented by but one species in our
northern waters. The characteristic of the order is
the tube-like structure of the different parts of the
frond. Each main stem branch or branchlet is a
single long undivided cell, filled with a green granular
substance, suspended in the watery fluids of the plant.

---

* Bryopsis = Moss-like

Bryopis plumosa,* Lam.

Perhaps the most beautiful of our green Algæ is the one here named. The artist gives, in Plate I., an admirable representation of a typical plant col lected by my friend Mr. A. R. Young, at Hell Gate, N. Y. The picture will give you a better idea of this interesting plant than any description in mere words. But it had better be said, that it commonly grows in tufts, a considerable number of fronds from the same point, from two to six inches high. The leading filament is beset all around, or sometimes on two opposite sides only, with long widely spreading branches, which are shorter toward the top of the plant. These, in their upper half, are clothed with long or short, straight branchlets, so placed as to give the plant a decidedly plumose or feathery appearance. It grows upon the rocks, or parasitical upon other Algæ, in shaded tide pools along our rocky shores. Mr. Collins informs me that it may be found upon the muddy bottoms of Mystic River, "where the tide ebbs and flows twice in twenty-four hours." I found some very beautiful specimens of it growing in a clear pool beside overhanging rocks on Ram Island, off the Marblehead

* Plumosa=feathery.

BRYOPSIS PLUMOSA. *Lam*

shore. Miss Booth found it floating up from deep
water at Orient, L. I. Mrs. Davis collects it in
tide pools at Gloucester. It is not a rare plant,
though not very common. It may be found from
July to October, and very likely later. I have some
very fine plants collected by Mr. Young, at Hell
Gate, New York city, the last part of September.
It may no doubt be looked for in the same situations on the Pacific coast, as it grows nearly all
over the globe. I have a fine specimen from Dr.
Dimmick, of Santa Barbara, California. It is of
a dark green color, and its delicate feathery frond
can never be mistaken, when seen displayed in all
its rare beauty in the crystal waters of the rocky
basins where it makes its home. When mounted
and dry it adheres well to paper and has a peculiar
glossy look.

Order.— *ZOOSPOREÆ.*
Genus — *ENTEROMORPHA,** Link.*

The plants of this genus are of a bright green
color, resemble the *Ulva* in structure, and grow in
much the same situations along side of that, and

* Enteromorpha = Intestine-shaped.

mingled with it in tide pools and upon the rocks
between tides. They are distinguished from that by
their *smaller* and *tubular* fronds. There are three
American species of this genus, common everywhere,
on both sides of the continent, and easily distinguished
from each other.

ENTEROMORPHA INTESTINALIS LINK.

The first named species is a simple unbranched
frond. Very slender at the bottom, it gradually
expands to the width of half an inch or more, sometimes an inch and a half, and grows from six to ten
inches high. It keeps nearly of the same width
throughout. When found growing in the tide pools,
it will usually be seen to be inflated, or filled with air
bubbles. Being filled out in this way, and at the
same time a little constricted at irregular intervals,
it has a decidedly intestinal appearance. The color
is a light green, but portions of the frond, especially
at the top, will often be found colorless or white,
owing to the fact that the chlorophyl, or green coloring
matter of the cells, has been discharged. The *unbranched inflated frond* distinguishes this species.

ENTEROMORPHA COMPRESSA GREV.

In this species the frond is compressed or flattened,
and is never inflated. The two layers of cells which

make up the substance of the frond appear never to
be separated. This is the most widely distributed of
the species of this genus. It is found in all waters
rom the equator to the arctic circle, and beyond. It
is extremely slender at the base, but gradually
expands upwards. The branches come out mostly
near the bottom, are themselves commonly unbranched,
and are neither so wide nor so long as the fronds of
the last species. They mostly have blunt tops which
look as though they had been cut square off. Most
of my plants are three or four inches high, though
I have some but an inch, and some quite eight inches.
The color is a little darker green than the last, and
the substance thicker. The *branched frond* distinguishes this species from the last, and the *simple
unbranched branches* distinguishes it from the next.

ENTEROMORPHA CLATHRATA, GREV.

This is by far the most variable of our *Enteromorpha.*
It is more slender than *E. compressa*, or any typical
form of *E. intestinalis.* It is often so fine and hair-
like, that you will certainly think it a *Cladophora.*
But a careful look at it with your pocket lens will
show you that the stem and branches are not made
up of a string of single cells, placed end to end, as
in that genus. This plant is profusely branched, and

the branches are divided and subdivided until they are no thicker at the ends than human hairs. The lesser branches are apt to be spiney. I have specimens of *E. clathrata* in my herbarium, whose fronds are nowhere more than one-eighth of an inch wide, though they are a foot and a half long. They will be found of various lengths, from two or three inches up. Under a high magnifying power, the cells composing the frond will be found to be quite square, and placed in a regular rectangular order, so that the frond will appear tesselated or latticed; hence its name.

Genus.— *ULVA,* L.*

The largest bright green plants in all seas belong to this genus. Two species are usually quite large when full grown, though there are plenty of them in the young state, and the collector will find them in abundance no more than two or three inches high. The first two species are common on both coasts; the last grows only on the Pacific.

ULVA LATISSIMA, L.

The *widest Ulva* is extremely variable in size and

* Ulva, from Ul = water in Celtic.

shape, varying in respect to the former from two to
twelve inches in width, and from six to twenty-four
and thirty-six inches in length. And in respect to
the latter, it is sometimes simple, and sometimes
lobed, sometimes plain, and as broad as long, sometimes long ruffled, or plaited on the edge. The
substance of the frond is thin and soft, and very
smooth and glossy, like silk. The color is a brilliant
green, being darker the deeper the water it grows
in. It sometimes turns brownish in the herbarium.
It is often found pierced with holes, the results either
of age or of the attack of snails. It is an annual,
but is often found in winter. It grows in pools and
below low-tide mark. It is so common everywhere
that I need not give special habitats.

Var. *Linza L.*—This is a charming and interesting
plant. Starting from a minute "hold-fast," as we
call the root, or place of attachment of the plant in
Algæ, it gradually expands to the breadth of an inch
or more, and rises to the height of six or eight inches.
The edges are full or ruffled, so that when
spread out on paper, the plant seems plaited all
down the sides, and the full grass green color of the
frond is deepened at every plait. Our figure, Plate
II., gives a very good account of it. It is quite
common along our rocky shores northward, adheres

well to paper, and is, by far, the most beautiful and most manageable of our *Ulvæ*, for the herbarium.

ULVA LACTUCA,[*] L.

The full grown plant differs from the polymorphus *latissima*, which it in most respects, much resembles, chiefly in these two particulars. It is of a paler color, and a much thinner substance. On dissection, it is found to consist of but one layer of cells, while *U. lattissima* has two layers. This fact, no doubt, accounts for both the peculiarities named above. When young, it is said to form an inflated bag like an overgrown *Enteromorpha intestinalis*, then at length by splitting along the side, floats out a thin membrane of but one layer of cells. It is an annual, and appears in spring and summer along with, but not so common as *U. lattissima*. I found it in August, very plentiful and very large at Southold, L. I.

ULVA FASCIATA,[†] DELILE.

The frond is more rigid even than that of *U. lattissima;* rises from a short stem, and is divided into several strap-shaped segments half to three-fourths of an inch wide, of nearly equal breadth throughout, and six

---

[*] Lactuca = lettuce.
[†] Fasciata = bundled.

V. LATISSIMA / ... U?? ... LINZA

or eight inches long, either simple or forked. The margin is mostly toothed and frequently undulate. The color is a full grass green, and the plants in my herbarium certainly keep their color much better than the *Ulva* of our coast. My plants adhere well to paper. It is found in abundance at Santa Barbara, California, but my correspondents do not elsewhere report it from that coast.

Genus.—*PORPHYRA,** *Ag.*

In structure, as well as in habit of growth, and method of reproduction, this Genus agrees very well with the *Ulva.* There is but one species in this genus.

PORPHYRA VULGARIS, AG. " LAVER."

Common everywhere. It is known by its frond of dark purple, thin and somewhat elastic membrane, which has a peculiar sheen like that of satin. This quality of it is retained somewhat even when dry, but is very striking and beautiful when the plant is in the water. The frond is as variable in form as that of the *Ulva*, from which it differs mainly in respect to color. I have often found it near low

* Porphyra = purple-weed.

tide, growing attached to boulder rocks, a great broad membrane, ten inches across, attached by a single point near the middle of the frond; again it will put forth a number of segments of such a frond, attached by their sides to one point; again a narrowish frond a foot long or more, attached by a short stem at one end. But the purple or brownish color, and the "sheeny" smooth, satin-like appearance of the frond will always serve to identify it. It is much used in Great Britain as an article of food for a relish with roast meat. The Chinese use it for making some sort of soup. The North Adams Colony imported it by barrels from China at one time. It does not adhere well to paper in drying, shrinking and pulling away. But it is said, that if the cloth is not removed from it at all till it has been under heavy pressure for a considerable time, and is fully dry, it adheres perfectly to the paper. It is an annual, and may be found the season through. I have fine specimens of it from California and from China, which have a rich dark purple color. And I have it from England as red as the "Dulse." But my plants from the shores of Massachusetts Bay are of a very decided brown.

Genus.—*CLADOPHORA,* Kütz.

No less than nineteen species of this genus are enumerated in Dr. Farlow's list of 1876, at least, fifteen of which are said to be natives of our northern shores. But our best botanists think the genus sadly in need of revision, for this country at least ; and assert that certainly two distinct systems of classification and nomenclature prevail in Europe. I shall attempt to give an account here of those species only which I believe can be so described as to be easily determined by the Amateur Collector. For the rest, you must needs make resort to the friendly aid of those botanists, whose ample suites of specimens will enable them, by comparison with yours, to determine your plant at a glance. The plants belonging to this genus, make up no inconsiderable portion of the green flora of our waters, and many of them make very beautiful specimens for the herbarium. The genus is characterized by extreme simplicity of structure. The main stem and branches alike consist of a sort of jointed thread, made up of single cells, attached end to end. The plants are always profusely branched, and in this regard are distinguished from those of the next genus, which are never branched.

* Cladophora = branch-bearing.

Cladophora arcta, Dillw.

The *arched Cladophora*, of which we give a fine
and characteristic illustration in Plate III., is named
from the peculiar habit of its growth. The branches
divide and subdivide by extremely acute angles, and
the ramifications are all very straight. This prevents
the unsymmetrical outline common to most plants of
this genus, keeps the branches somewhat close together
as they rise upward, and, at the same time, permits
them to separate gradually and symmetrically. This
gives the tuft its arched and graceful form, not unlike
the outline of our more perfect and beautiful elms.
This characteristic of form, the yellowish green color,
and the decidedly glossy or silky look, which the
plant usually presents when dry and pressed on paper,
makes its determination easy. Another peculiarity
which may be noticed in the dried specimen is the
disposition of the chlorophyl of the terminal branchlets
to collect in the extreme end cell, making that cell
have a distinctly darker green color than the cells just
below it in the branch. It is an annual. Mr. Collins
finds it common at Nahant and Nantasket, on rocks
between tides, from March to July. Miss Booth finds it
extremely rare at Peconic Bay, L. I. At Marblehead I
gathered it frequently during the summer months. It is
often found on the California coast, near Santa Cruz.

CLADOPHORA UNCIALIS, FL. DAN.

As its name implies, is about an inch long. I have found it growing in tide pools, or on the rocks near low tide, in little globose tufts, about an inch across, and of the same height. The tuft grows from a mass of matted root-fibres. It is more or less closely matted together by reason of its wide and irregular branching. When growing, the plant is of a bright green color, which will be discharged if it is put into fresh water. When dry it is quite a yellowish green, lighter still toward the centre of the tuft. The cells of the main stems and branches are of nearly uniform length, and two or three times longer than broad. My plants are all from Marblehead where they were collected in midsummer. Mr. Collins finds this plant in the same localities, seasons and situations, as the *C. arcta*, which it resembles not a little. My other correspondents do not report it, though no doubt it may be found along our whole northern coast.

CLADOPHORA RUPESTRIS, L.

The *Cladophora* " of the rock," is a very distinctly marked species. It grows between tides and below. Its best forms are to be found in tide pools near low water mark, or on the perpendicular sides of rocks,

near low tide, under the curtain of the overhanging
Fuci. It is a very dark, dull green. Its filaments are
coarse, stiff, straight and rigid. Its secondary branches
divide at very acute angles, and therefore, as in *C.
arcta*, cluster and cling somewhat closely about the
principal branches. There is a decided tendency in
the main branches to separate from each other, and
stand aloof with their closely clustering branchlets.
These separate pencils of dark green filaments are of
quite unequal length. The tuft is commonly three
or four inches high, but sometimes, six or eight. It
is not uncommon from New York city northward; but
it certainly is more beautiful on our northern New
England shores. It is reported from Nahant and
Cape Ann, by Mr. Collins and Mrs. Bray, from March
to December.

CLADOPHORA CARTILAGINEA, RUPR.

Is a California plant, and is found growing, as Dr.
Anderson informs me, at all seasons, on rocks and
other sea weeds, in tide pools, very common at Santa
Cruz. Its robust, coarse frond; perceptable harsh-
ness to the touch; dull green color; stiff, straight
branches, set at an acute angle with the stem; its
refusal to adhere to the paper, as well as its general
appearance, relate it closely with *C. rupestris.* It

differs in being of a shade lighter color, and a somewhat slenderer filament. This is almost the only
Cladophora which gets sent over here from California,
though it is not the only one growing there. It is
reported common all along the coast.

CLADOPHORA REFRACTA,* ROTH.

This plant grows on rocky shores in tide pools.
The filaments are very slender and fine, profusely
branched. The end branchlets are so profuse, and
so widely set, even recurved, or bent back, that they
give the plant a very decidedly feathery, or downy
appearance all along the edges of the frond and
branches. This is its most characteristic mark. It
is a bright green in the water, but fades a good deal
when dried and mounted. It grows three or four
inches high. It is a summer annual, and may be
looked for on the whole coast, in tide pools, or floating up from deep water.

CLADOPHORA GRACILIS,† GRIFF.

This species grows in deep water, parasitical upon
Zostera and smaller Algæ in the *Laminaria* region.
It generally has its main branches much interwoven

* Refracta = bent back.
† Gracilis = slender, graceful.

and entangled, so that it will look like a formless
mass of green as it rises to the surface of the water
and washes on shore. The only guiding mark is its
long, straight, or inwardly curved ultimate branchlets.
These are conspicuous, and the cells of which they
are made are also seven or eight times longer than
broad. The filaments are as fine as human hair, six
or eight inches long, and have a very silky look when
massed in the mounted specimen. The color is a
very bright yellowish green when fresh. Mr. Collins
finds it at Nahant between tide marks. It is a summer
plant.

Cladophora glaucescens, Griff.

Grows in tufts not much entangled, on stones and
rocks, between tide marks and in pools, from three to
five inches high. The branches divide and subdivide
excessively, are quite slender, and the ultimate branches
are closely beset usually on the inside, almost always
on one side only, with a series of straight, acutely
branching undivided branchlets, composed of several
cells. In drying, the chlorophyl is usually dissipated
to one end of the cell, making the plant under the
lens look somewhat variegated. The filaments are
constricted at the joints of the cells. Color a pale or
glaucous green.

CLADOPHORA FLUXUOSA, GRIFF.

Harvey Considers this plant nearly related to the last, if it is even specifically distinct. It is chiefly distinguished by its less compound habit, the length and nakedness of the principal branches, and their fluxuosity. It grows in rock pools between tides, is not very common, and is found both north and south of Cape Cod.

CLADOPHORA LÆTEVIRENS, DILLW.

The filaments are rather loosely tufted, feathery, robust and somewhat firm or rigid; color, a pale green, as its name indicates, faded, and without gloss when dry. "Filaments three to four inches long, or more, much branched, main stem flexuous or angularly bent, set with alternate or scattered occasionally opposite, repeatedly decompound patent branches." Articulations of the main stem, four to eight times, of the ramuli, three to four times as long as broad. Substance not very soft. It adheres, but not very strongly, to paper, in drying. It is found in New York Bay, on the Massachusetts coast, and in California, in the latter region being quite common. Mr. Collins has collected it at Nahant and Revere between tide marks.

Genus.— *CHÆTOMORPHA*,* *Kütz.*

The plants of this genus may be separated into
two groups, the straight and the crooked. The first
we shall commonly find growing in their native haunts,
standing up straight, stark and rigid. The others we
shall find usually floating, or thrown on shore among
the sea weed, a twisted, matted, entangled mass of
long green threads, thick or slender, and as crooked
and kinked as wool. The plants of this genus
consist in general, of a single long, bristly, jointed,
unbranched, green thread.

CHÆTOMORPHA MELAGONIUM, WEB. & MOHR.

This species grows in rock pools near low-water
and below. From a disk-shaped root, on the rock,
it rises up four to twelve inches, solitary, straight, stiff
and wirey, of a dark green color, as its name signifies,
twice as thick as a bristle, tapering to the base, and
blunt at the top. Articulations two or three times
longer than broad. Common all along our rocky
shores north of Boston, from June to October.

CHÆTOMORPHA ÆREA, DILLW.

This plant has something the same habit as the
last. It grows in the same situations along the whole

* Chætomorpha = like a horse's mane.

coast; but more common south of Cape Cod. It is common in southern California. It is but half the thickness of the other, and is not nearly so stiff and rigid, and grows not solitary, but in tufts, from three to twelve inches long. The filaments are considerably constricted at the joints. The articulations are about as broad as long. The color is yellow green, fading in the herbarium, and turning darker. Young plants are straight, but the old ones are often bent. It does not readily adhere to paper.

CHÆTORMORPHA OLNEYI, HARV.

Filaments in tufts, about the size of the last, as thick as a bristle, straight or bent, or much contorted; pale green; articulations once and a half times longer than broad. It is of a much softer substance than the last, though it feels harsh when dried on paper, to which it adheres firmly. I found it beyond the first beach at Newport, Aug. 7, much contorted, like *C. Picquotiana*. It was named for Mr. S. T. Olney, of Providence.

CHÆTORMORPHA PICQUOTIANA, MONT.

Filaments loosely bundled together in masses; grass green; rigid, glossy, twelve inches long or more, twice as thick as bristles, variously curved and twisted; articulations three to five times as long as broad

5

constricted at the joints. In drying, the plant fades a little, but keeps its glossy look, and as the chlorophyl collects at the ends of the cells it gets a variegated appearance, an alternation of light and dark points along the thread. It is common along the whole coast. It grows in deep water five or six fathoms down, and so must be sought for among the cast up sea weeds, or floating on the surface of the water. Mr. Collins found it in tide pools, at Revere, in the spring, but it may be found all summer. It does not adhere to paper.

CHÆTORMORPHA TORTUOSA, DILLW.

You will find upon the rocks, or upon the Algæ growing on them, mats of green wool, spread out or rolled up. This is *C. tortuosa*. Its filaments are very fine, finer than human hair, densely interwoven and felted together into rolls, or spreading mats. It does not colapse when taken from the water. It is common at Nahant, Marblehead, and Nantasket, and northward in midsummer. My specimens have adhered very well to paper. It is not uncommon in California.

OLIVE GREEN ALGÆ.

I. FROND LEAF-BEARING.

Main stem and branches cylindrical, bearing globular, stalked minute, air vessels, and narrow, undivided, dotted leaves. General habit arborescent. "Gulf-weed." *Sargassum*.

II. FROND, FLAT, CORIACEOUS OR LEATHERY.

1. *With Midrib*.

(*a*.) Frond perforated. *Agarum*.

(*b*.) Frond entire, stem bearing leaflets or wings. *Alaria*.

2. *Without Midrib*.

(*a*.) Frond thick, leathery and large, dark olive green or brown. "Kelp." *Laminaria*.

(*b*.) Frond thinner and smaller, light green or brown, from three to twelve inches long. *Punctaria*.

(*c*.) Frond narrow in proportion to length, half-inch wide, eight to twelve inches long. *Phyllitis*.

(*d*.) Frond still narrower and constricted at intervals. *Scytosiphon*.

III. Frond narrow, compressed or flattened.

Frond forked or branched, thick, tough, one to two feet long. " Rockweed." *Fucus.*

IV. Frond filiform or thread-like.

1. Frond *Unbranched.*

Frond four to six inches long. Sometimes inflated and constricted; always covered with minute dark dots. Color, yellow olive. *Asperococcus.*

Long, ten to twenty feet, elastic, much attenuated at each end. *Chorda.*

2. Frond *Branched.*

(*a.*) Branches mostly simple.

Long in proportion to main stem, parts as large as pack-thread. Color black. *Chordaria.*

Short in proportion to main stem. Color, olive or full green. *Castagnea.*

(*b.*) Branches, naked, divided and sub-divided.

Stem and branches repeatedly forking. Color, yellowish olive, dotted over with minute dark colored warts, frond six inches high. *Stilophora.*

Frond one to two feet long, intricately branched; branches at last very small. *Dictyosiphon.*

(*c.*)   Branches clothed :

   1.   With rows or circles of closely set, very short spines, which overlap each other, thus covering every part of the frond.

Cladostephus.

   2.   With short, fine, light olive green, delicate fibrils, which fall away and leave bare spines ; or with long darker green pencils of hair-like filaments.   *Desmarestia.*

V.   FROND CAPILLARY.

   1.   *Unbranched.*

Frond small, parasitical on *Fucus*, tufted.

Elachista.

   2.   *Branched.*

Frond fine, profusely branched ; from a yellowish to a bright green ; parasitical on *Fucus, Chorda, Chordaria* and other Algæ.   *Ectocarpus.*

VI.   FROND TUBERFORM.

Fronds look not unlike green tomatoes.

Leathesia.

KEY TO THE GENERA OF THE PACIFIC COAST.*

I. FROND, LEAF-BEARING.

1. Stem flattened, rough, leaves on each edge, air vessels in the stems of some of the leaves; plant many feet, sometimes many yards long. *Phyllospora.*

2. Plant from a few to several hundred feet long. Stem cylindrical, slender, branched, leaves on opposite sides of the stem. Air vessel in each leaf stalk. Root, large, much branched. *Macrocystis*

3. Stem long, slender, cylindrical, elastic, terminated in a large rounded air vessel which is crowned with a large tuft of long, slender leaves. Root branched. *Nereocystis.*

4. Stem short, stout, cylindrical, surmounted at top with a large tuft of deeply ribbed leaves. *Postelsia.*

*Only those genera which have species peculiar to this coast are included in this Key, all the rest are in the other.

II. Frond flat, leathery.

    1. Stem long or short, mostly slender. Blade thick, leathery, large or small, dark olive green or brown. "Kelp."

                    Laminaria.

    2. Stem cylindrical, long, stout, winged on each side with long stalked, leathery leaflets. Blade of frond thick, long; midrib at base, which fades out towards the top.   *Pterygophora.*

    3. Stem short, split, blade long, covered with a net-work of prominent nerves.

                    Dictyoneuron.

III. Frond flattened.

    1. Frond narrow, thick, tough, forked, from three inches to two feet long. "Rock-weed."   *Fucus.*

    2. Frond leafy below, finely divided and filiform above. Air vessels in the swollen bead-like ultimate branchlets. *Halidrys.*

    3. Frond flat, narrow, thin, pinnately compounded, pinnæ and pinnulæ tapering to top and bottom.   *Desmarestia.*

    4. Frond flat, fan-shaped, small, marked with concentric zones or belts of darker color.   *Zonaria.*

IV. FROND CYLINDRICAL, FILIFORM.

    Frond branched from leading stem, branches short, thick as pack-thread. Plant four to ten inches high. Color black.

                            Chordaria.

V. FROND TUBERFORM.

    Frond inflated, massed, thin and soft, yellow olive, from one to three inches through. *Asperococcus sinuosus.*

CHAPTER III.

OLIVE COLORED ALGÆ.

Down on the shore, on the sunny shore!
  Where the salt smell cheers the land;
Where the tide moves bright under boundless light,
  And the surge on the glittering strand;
Where the children wade in the shallow pools,
  Or run from the path in play ;
With the hushing waves on its golden floor
  To sing a tuneful roundelay.
Down on the shore, on the stormy shore!
  Beset by growling sea,
Whose mad waves leap on the rocky steep,
  Like wolves up a traveller's tree,
Where the foam flies wide, and an angry blast
  Blows the curlew off with a screech ;
Where the brown sea-wrack, torn up by the roots,
  Is flung out of fishers' reach .
Where the tall ship rolls on the hidden shoals,
  And scatter her planks on the beach.

CHAPTER III.

Sub-class.— *MELANOSPORÆ.*

Order.—*DICTYOTEÆ.*

Genus.— *ZONARIA,** *Ag.*

ZONARIA TOURNEFORTII, LAM.

MANY plants of this species have been dis-tributed under the name of *Z. flava.* It is common in southern California, as some species of this genus are in all tropical and sub-tropical seas. It grows from a short, flattened stem, a widely-spreading, flat, fan-shaped frond, two to four inches high, with obscure concentric bands of a darker color on the olive green of the plant. The extreme rounded

* Zonaria — belted or zoned.

thin edges of the lobes are bordered with a fine dark line. The frond is split down from the margin with clefts running down quite to the base, or half-way or a quarter of the way, and the lobes are more or less profusely sprinkled over with dark colored fruit dots. It may be found throughout the season at Santa Barbara and San Diego, upon small rocks near low tide, or thrown up by storms upon the beach.

Order.—*FUCACEÆ.*
Genus —*SARGASSUM,* Ag.

This genus is represented by but one species on our north Atlantic coast. But this species is common enough along most of the shores south of Cape Cod.

SARGASSUM VULGARE, AG.

The plant grows from a flat discoid hold-fast, with a filiform stem as thick as stout wrapping-twine, which branches alternately, and bears on the main stem and branches long narrow leaves, which have stalks or petioles, a well-defined midrib and toothed edges, and are marked on the surface with

* Sargassum, from Sargazo, Spanish for Sea-lentils.

minute dark dots. The leaves vary greatly in length and breadth and even in shape, being from one to three inches long, and from one-eighth to one-third of an inch wide. The air vessels which distinguish the genus are numerous little globes, one-eighth of an inch or more in diameter, set upon little stalks half an inch long, which grow from the axils of the leaves. Sometimes from the appearance of a sharp tip or point on the opposite side of the globes, the stalk seems to extend quite through it. The fruit is borne in a many times branched "twiggy," thickened receptacle, which grows from the axils of the leaves. I have found this plant growing common upon small stones and pebbles all along our southern New England coast, just below low-tide marks, usually less than two feet long, though I have plants not less than four feet. But the length will depend mostly upon the age. Plants not more than a foot long make the best herbarium specimens. It is perennial.

Genus.— *PHYLLOSPORA,* * *Ag.*

PHYLLOSPORA MENZIESII, AG.

This is a very common plant, growing along the whole California coast, at all seasons, upon rocks between tides and below. It is found on the sea beach of the ocean and Bay, at San Diego, thrown up from deep water, and at Castle Point, Santa Barbara, in deep water. From a branching hold-fast, a short, round stem rises, which immediately divides irregularly, into several long, flattened strap-like branches, many feet, sometimes many fathoms long, from one-quarter to one inch wide, thickish, roughened, or smooth, and bordered on each edge with a profusion of leaves. The leaves are wide and rounded at top, narrow or distinctly stalked at bottom, varying in length from one-half inch to six or more inches. Sometimes set an inch apart, sometimes crowded close together, and interspersed at intervals with large, pear-shaped air vessels, one-half to three-quarters of an inch in diameter, these are often tipped with a leaflet. The plant may be infallibly determined by the distinctive marks given above. It should be partly dried before putting in the press.

* Phyllospora = Spore-bearing leaves.

Genus. — *HALIDRYS,** *Lyngb.*

HALIDRYS OSMUNDACEA, HARV.

This elegant plant forms a prominent feature in the marine flora of southern California. It grows in abundance at San Diego, below tide, and in the sluice-ways cut in the rocks by the water. It is thrown on shore at all seasons. It is also abundant at Santa Barbara, but absent at Santa Cruz. At all events, that acute observer, Dr. Anderson, does not report it as present. It grows from a discoid hold-fast, a roundish flattened stem, as thick as a goose quill. Flattening more and more upwards, the stem divides or branches, and puts out from its edges, winglets, or alternate leaves, from one to two inches long, which, like the flattened stem, are thick and midribed. Near the middle of the stem these cease, and the stem becomes rounded again and alternately branched, the branches also branching alternately in nearly the same plane. The secondary cylindrical branchlets form the air vessels of the plant, by being much swollen and hollow, and constricted at regular intervals, giving them an appearance not unlike a string of coarse black beads.

* Halidrys = Sea Oak.

The full grown plant must be two or three feet long, though my specimens do not show it. It is olive green when fresh, but like most of the *Fucaceæ* turns black in drying.

Genus.— *FUCUS,** L.*

The plants of this genus are together popularly known as " Rockweed." They constitute, on the Atlantic coast at least, more than one-half of the mass of our littoral Algæ. There are three species sufficiently common on the Atlantic coast to come within the scope of this book, and one on the Pacific. The latter will be described first, it standing thus in the natural order.

FUCUS FASTIGATUS, AG.

This species seems to be the most common *Fucus* in southern California, though *F. vesiculosus* grows there in abundance, as it does also along the coast north ; and *F. Harveyanus* is found as a rare plant at Santa Barbara, and as a common one at Monterey. Mr. Cleveland says that *F. fastigiatus* grows at San Diego in mats, on flat rocks left uncovered by the ebb tide, at all seasons, abundant.

* Fucus = Seaweed.

It has a cylindrical frond as thick as a sparrow's quill, which forks very near the base, and again each of the parts repeatedly fork more and more remotely, but less and less widely, six or seven times. The fruit is borne in the thickened terminal branchlets. It grows to the height of three or four inches. There are no air vessels.

FUCUS VESICULOSUS, L. —"ROCKWEED." ✓

This is the *Fucus* with little bladders, or air vessels. Of the two *Fuci* which cover the rocks and wood-work of wharves, along our whole eastern coast, as far south as the Carolinas, the most plentiful is the one named above. This and the next, grow together everywhere. The plants of this species are greatly variable in size according to their place of growth, being most luxuriant where they have the tide longest. The frond varies from a quarter of an inch to one and one-half inches in width, and from two inches to two feet in length. It is tough and leathery in substance, decidedly flat, with an evident midrib throughout the main stem and branches. It branches by forking, and the axils of the divisions are usually very acute. Each frond is commonly provided with from one to several pairs of oval air bladders, immersed in the substance of the frond, each side of the midrib. It bears its seed vessels in the extremities of the branches, which are, in that case, much swollen, and of a pronounced yellow color. Cut through with a knife, these swollen receptacles will appear to consist of a mass of hard gelatine, and the seed vessels will show themselves as bright yellow spots, all around the circumference. The distinct olive green color of the fresh plant changes to black in drying.

FUCUS NODOSUS, L.

Our next most common " Rockweed," is the "knotty " *Fucus*, so called, from the knots or swellings which the interior air vessels make in the frond. This species differs from the last in several important respects : first, by having a very narrow frond, of the same width throughout, one-quarter of an inch or more ; second, by its method of branching, which is not in regular forks, but by putting out side branches of various and irregular lengths, commonly quite long, from the sides of the main stem ; third, by the presence also with the branches of short (three-quarters to one inch long) branchlets, whose wider ends thicken and produce the seed vessels ; and fourth, by the prominent swellings or knots in the stem, and branches which give the species its name. This and the other *Fuci* are fastened to the rock on which they grow by a discoid hold-fast. The plants grow between tides from six inches to two feet long. It is a perennial, and the old fronds will be quite likely to have some species of *Ectocarpus* growing on them. It is also the favorite and almost the only home of the *Polysiphonia fastigiata*. It is a rich olive in water, but quite black when dry.

FUCUS FURCATUS, AG.

The *forked Fucus* resembles the *F. vesiculosus* in its general habit of growth, but differs from it in several particulars, viz., in having a somewhat wider, shorter and more constantly typical frond, in having no air bladders, and in having the terminal forks which bear the seed vessels much longer, more pointed, and less swollen, being two and sometimes three inches long. The whole plant is a foot or more in length, and grows just down at the extreme low-water mark. It may be most easily found and collected, during the time of "spring tides," at new or full moon. It is common on the rocks at Nahant, Marblehead, and northward. The microscopist distinguishes this species from *F. vesiculosus* by a difference in the contents of the seed vessels. There are two other species of *Fucus* recognized in our north eastern flora. *F. ceranoides* at Marblehead, and *F.*

serratus at Newburyport ; but their rarity makes it
undesirable to describe them in a work intended only
as a popular introduction to the more common forms
of our marine flora.

Order.— *PILÆOSPOREÆ.*
Sub-order.— *LAMINARIEÆ.*
Genus.— *MACROCYSTIS,* *Ag.*

MACROCYSTIS PYRIFERA, AG.

This is the giant among sea weeds. Indeed, it
attains a length unknown in any other vegetable form
upon the globe. Were it not to question the testi-
mony of careful observers, I should be much inclined
to doubt some of the stories told about this remarkable
plant. Dr. Hooker says it attains a length of 700
feet, and Bory St. Vincent declares it is sometimes
found 1,500 feet long. Mrs. Bingham, of Santa
Barbara, writes me that it is frequently thrown on
shore there, 100 feet long. Mr. Cleveland, who has
been at great pains and trouble to get me exact
data as well as typical specimens of this plant, has

* Macrocystis = With large bladders.

seen it 200 feet long at San Diego. The account
which I give is from their notes. The hold-fast for
these larger plants is a great mass of branching roots.
"as large as a bushel basket," sometimes three feet
broad, and a foot thick, which cling to the rocks and
boulders with great tenacity. One or more stems, from
a half to three-fourths of an inch in diameter rise from
this, putting out leaves on either side alternately, a
foot apart at the base, gradually growing nearer toward
the end of the stem. The leaves, in the largest plants,
are from two to four feet long, and three or four
inches wide, stalked, and the stalk swollen into a pear-
shaped air vessel, sometimes an inch and a half long,
and an inch thick. The leaves are thin, peculiarly
wrinkled, of a fine olive color, and along both edges
bordered with sharp, spine-like teeth, which point
forward. These plants grow in water, fifty feet deep
or more, in vast forests, coming to the surface
and then stretching their leafy fronds far out, prone
upon the sea. In this way, great fields of them,
sometimes a mile wide and several miles long, are
formed, especially near bays, as at San Luis Obespo,
Santa Barbara, San Pedro, and San Diego. The
stem terminates in a leaf-like expansion, and the
growth goes forward in a very curious fashion, by the
constant splitting off of the side of this terminal leaf.

The splitting is a natural process, and as it proceeds, the petiole and the air vessel are successively developed, so that when the tip of the leaflet, finally parts from the parent leaf, it will be fully formed, though not full grown. At the same time there will be lying inside of this four or five other leaflets, in various stages of growth, from the most rudimentary, to the almost fully formed. I suppose this must be considered the most remarkable feature of the marine flora of the Pacific coast, though it is by no means the only wonderful plant that makes its home in those waters.

Genus.— *NEREOCYSTIS,*[*] *Post. & Rupr.*

NEREOCYSTIS LÜTKEANA, POST. & RUPR.

Next to the *Macrocystis,* the *Nereocystis* is the most remarkable and wonderful plant of the Pacific waters. To quote Harvey, "The *Nereocystis* of the North West coast, is said, when fully grown, to have a stem measuring 300 feet in length, which bears at its summit a huge air vessel, six or eight feet long, shaped like a great cask, and ending in a tuft of upwards of fifty forked leaves, each of which is from

* Nereocystis = Sea-bladder.

thirty to forty feet in length. The cask-like air vessel which may be eight inches or more in diameter, buoys up this immense frond, which like Milton's hero, lies

' Prone on the flood extended long and large,
Floating many a rood.'

Here the Sea Otter has his favorite lair, resting himself on the vesicle, or hiding among the leaves while he pursues his fishing. The stem which anchors this floating mass of fronds is of considerable length and elasticity, though it is no larger than a whip cord. It is employed as a fishing line by the rude natives of the coast."

Dr. Anderson, of Santa Cruz, was kind enough to send me a small typical specimen, sufficiently large to show all the characteristic points in the form and growth of the younger plants. Starting from a many-pronged hold-fast, like that of the *Laminaria*, is a slender stem not more than a quarter of an inch in diameter. For two yards it keeps this size, when it begins to expand. For the space of another yard it gradually increases in size, and is evidently hollow, till at the end it has attained a diameter of one and a quarter inches, when dry; it probably was something more than that in the water. Then it is immediately and suddenly drawn in, or constricted, and forms a narrow neck,

not more than three-quarters of an inch through,
and then as suddenly expands into a large, egg-
shaped vesicle, the narrow end of the egg being
next to the neck, and the wide end crowned with
two tufts of long, narrow leaves. The dimensions
of the oviform part of the air vessel are, in the
long diameter two and three-quarters inches, and in
the short two and a quarter inches. The leaves are
from one-half a yard to a yard long and from half
an inch to one inch wide, many of them with thick
brown patches of spores upon them.

Mr. Cleveland has had the kindness to send me
parts of a plant and drawings of the whole, which
enables me to add a point to the history of this
curious genus, that I think will be interesting to
collectors. This form differs from the one already
described, by the air vessel bearing upon its apex
a single large forking petiole, whose two arms spread
out on each side and branch, like the antlers of a
deer; each short "prong" bearing, at the end, a
broad, long leaf. In a plant whose air vessel measures
5 ½ inches in diameter, the flattened petiole at base
was two inches broad, and the two "horns" into which
it immediately divided, were 1 ¼ inches broad and eight
feet long. These gave out branches upon the inside at
intervals of about a foot, which branches, at a distance

Dana

from their base of a foot or so, forked, and bore on each part a long, broad tongue-shaped leaf, two or three feet long, and as many inches broad.

Prof. Eaton has kindly sent me a copy of Areschoug's description (in *Botaniska Notiser* for May 15, 1876), of what he, with some hesitation, names a new species: *N. gigantea*, which answers very well to Mr. Cleveland's plant. It would seem to be an easy matter for our California botanists to settle the question of whether or not these two extreme forms are always distinct, or insensibly pass into each other, in a large group of specimens; or whether the first is but the young of which the last is the mature form, as some botanists seem to think. Mr. Cleveland assures me that the last described form is quite constant.

It is a very common plant, growing in deep water, all along the west coast, at all seasons, and is flung on shore in great quantities by the storms.

Genus.— *POSTELSIA, Rupr.*

POSTELSIA* PALMÆFORMIS,† RUPR.

This species is quite common on the west coast

---

* Postelsia, named for A. Postels, a fellow-botanist with Ruprecht.

† Palmæformis = Palm-shaped.

from Santa Cruz northward.   I have seen but one
specimen of this curious and interesting plant, and
that was kindly sent me by Dr. Anderson.   It is a
small but apparently a typical one.   The excellent
figure and description given by Ruprecht leaves nothing in that line to be desired.   The main stem is
many pronged at the base, hollow, about half an
inch thick, which size is uniform, except that it tapers
a little near the top, and about a foot long.   It is
crowned with a cluster of stalked leaves a foot or
more long, an inch or so wide at the middle, tapering
to a point at the top, and set in pairs upon the long
forked petiole.   The leaves are curiously ribbed or
"fluted" lengthwise, the higher ribs being in the
middle.   An examination shows that the depressions
on one side correspond to the elevations on the other
side of the leaf.   It is found at all seasons on exposed
points, growing upon the rocks.

Genus.— *PTERYGOPHORA*,* *Rupr.*

PTERYGOPHORA CALIFORNICA, RUPR.

For a fine plant of this species I am also indebted
to the liberality of Dr. Anderson, and for a full

* Pterygophora = Wing-bearing.

account of its habits to the celebrated botanist who
has done so much to illustrate the marine flora of
the North Pacific, Dr. Ruprecht.

This plant more nearly approaches the *Alaria* than
any other of the *Laminarieæ*. Fastened to the rock
by a multitude of prongs which radiate from the base
of the stem, the stem itself rises two or three feet,
half an inch thick, mostly quite cylindrical, but flattened
near the top, where it gives off the characteristic " wings "
on each side. The " blade," or the main leaf, is two
feet or more long, three inches broad in the widest
part, frayed out at the top, and thickened through
the whole length in the middle with a midrib, which
is apparently a continuation of the stem. This midrib has not the definite outline which it has in the
Alaria, but is only a thickening of the middle of
the leaf which vanishes imperceptibly towards the
edges and the top. The " wings " are stalked, not
crowded close together as in the *Alaria*, but set in
pairs, some distance apart, along the opposite sides
of the main stem, four or five or more pairs of them,
from one to two feet long, and from one to one and
one-half inches broad, with no trace of a midrib.
Mr. Cleveland reports this plant common from February
to May, growing in deep water, along the coast as far
south as San Diego. Dr. Anderson finds it among

the commonest plants growing with the other *Laminarieæ* throughout the season at Santa Cruz, California.

Genus.—*ALARIA,** *Grev.*

ALARIA ESCULENTA, GREV.

The *edible Alaria* grows upon submerged rocks just below tide. It is a plant whose peculiar aspect makes it very easy of recognition and quite impossible to confound with any other. Unlike any other of the " Kelps," except the *Agarum* it has a stout midrib running the whole length of the plant. This together with the little cluster of ribless leaflets or wings, borne on each side of the stem, just below the blade, makes the plant absolutely distinct. These leaflets bear the spores or fruit, and are always present except on young plants. The plant makes its anchorage upon the rock by the same means as the *Laminarieæ* generally. The stem is from three inches to a foot long, cylindrical. The blade consists of a thin wavy, or ruffled olive colored membrane, from one to four inches wide,

* Alaria = winged.

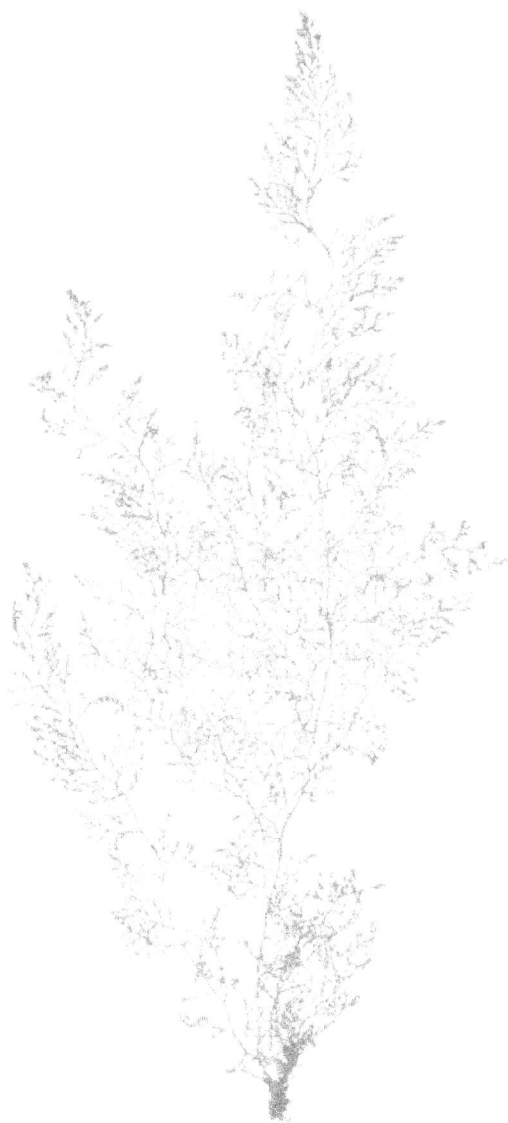

POLYSIPHONIA VIOLACEA, *Grev.*

developed on each side of the thick midrib. It
is of a delicate, tender texture, which easily tears,
and then always in the same definite oblique direction
toward the midrib. The ends of the old plants are
usually frayed out, the midrib protruding beyond
the rest of the blade with the "rags and tatters" of
the thin membrane hanging to it. The young plants,
when not more than six or eight inches high, make
very beautiful specimens, if neatly mounted. They
are of a very delicate green color, and adhere well
to paper, as, indeed, do my full grown plants. The
species is said to grow twenty feet or more in length
in some places. I have never found it over five or
six. On the outside of Ram Island, off the Marblehead shore, in midsummer, I found the rocks literally
covered with these interesting plants; and as they
hung out over the edge of the submerged cliffs, and
waved their long, delicate olive streamers in the green
rolling waters, they certainly presented a bit of sub-
marine scenery, well worth the trouble to find and
look at. Turner says that in his day, the midribs of
this plant stripped of the membrane, and the thickened,
fruit laden leaflets, were brought to market and sold
in Scotland, to eat, and were said to be sweet to the
taste. They are popularly called " Daber Locks." Mrs.
Bray finds it at Kittle Island and Magnolia on Cape

Ann, growing sometimes in tide pools. It need not be looked for south of Cape Cod.

Genus.—*DICTYONEURON*,* *Rupr.*

DICTYONEURON CALIFORNICUM, RUPR.

This is certainly one of the most interesting plants of this group. It was first brought from the coast of California, in 1840, by Wosnessenski, a Russian navigator, and described by Ruprecht. In addition to his excellent figure and full text, I have several specimens kindly sent me by Dr. Anderson, as a guide in giving an account of the plant. The one before me is about thirty inches long and two and three-fourths inches wide in the widest place, tapering somewhat toward the broken top, and rapidly to the stem below. The frond has a tendency to bend in the direction of one edge like a sabre blade. Its distinguishing mark consists, however, in the fact that both surfaces of the frond are woven over with a net-work of prominent veins and ribs, some of which run in a general direction, parallel with the edges of the frond, and others not so thick

* Dictyoneuron = Netted nerves.

or prominent, connect these in an irregular way, so that the "meshes" are of very indefinite size and shape. The hold-fast is a small bunch of branching roots, and the stem, which is flat, almost immediately expands into the blade. In most of the fronds, especially the older ones, the stem is split into halves, the split extending sometimes several inches into the blade of the frond. This splitting is a natural process, and not accidental. No collector of California Algæ ought to miss this curious and quite unique species. It may be found at Santa Cruz and northward, from June to November, among the other *Laminarieæ.*

Genus.— *LAMINARIA,** *Lam.*

The larger plants of this genus bear collectively several popular names, as "Kelp," "Oar Weed," "Devil's Aprons," etc. They are the largest Algæ belonging to the flora of our Atlantic coast. The three most common species to be named below, from that flora, may be easily distinguished from each other by well marked specific differences.

They are all deep water plants, and while they

Laminaria = A leaf.

would not be chosen for their beauty in the herbarium, they are certainly in the water, extremely graceful and interesting forms. They are all perennial. The method of drying, pressing and mounting them, has already been given in the Introductory Chapter.

Laminaria saccarhina, Lam.

This species is so named for the supposed sweet taste of the frond, a quality which I confess has thus far quite eluded my powers of detection. It is distinguished from the next species to be named, by its short stem, and its narrower frond. The stem is not more than four to eight inches long, and from one-third to one-half an inch thick. The stem terminates below in a conical mass of stout, root-like prongs, which constitute the hold-fast. These are firmly glued to whatever the plant grows upon, as shells, rocks, stones, etc., at the bottom of the sea. If you try to remove one of these large plants from its native anchorage, you will find that it holds very fast. The short stem expands upward abruptly, into a wide, thick, leathery, smooth, dark olive colored blade, eight to twelve inches wide, and six to eight feet long. It is usually wavy or ruffled at the edges. A narrow and very beautiful variety of this species grows along the shore at Newport, over by the beaches. It

is not more than three or four inches wide, but at
least two yards long. The frond is very smooth and
glossy, and exquisitely ruffled, so that as it rises and
falls with the undulating waters, like a streamer in the
upper air, it is, indeed, in form and motion, a thing
of rare grace. These plants lose most of their beauty
when dried and made ready for the herbarium. But
in the water they are most wonderfully fine. I want
to say a word for them because I know they are commonly either passed by without notice or countenence,
and rejected for their imputed ugliness. But you want
to see them at home if you would appreciate what
they may be under favoring conditions. To those
who make their summer home on Cape Ann, and
desire to see the wider forms of this species, as they
display themselves at their best, I would suggest that
you go along the rocky shore south of the village of
Rockport, out towards the Light House. As you come
near the end of the land, you will find many large
and deep tide pools, where these plants grow to perfection. There, as they bend with their wavy fronds
in long, graceful curves, over-arching the smaller Algæ,
which carpet the bottom, and decorate the sides of
the pool; their own rich olive brown color setting off
the brilliant reds and the bright greens of the other
plants; they do, indeed, help to make a picture of

exquisite beauty. This plant is very common on the
Atlantic coast north of New York city, also on the
Pacific.

LAMINARIA LONGICRURIS, DE LA PYL. √

The *long stemed Laminaria* is a plant which in our
New England waters grows to about the size of *L.
saccarhina*, except as to the stem which is usually
quite as long as the blade of the plant. The whole,
therefore, is from twelve to sixteen feet long, and I
have found it at Marblehead eighteen to twenty feet
long, the blade twelve to sixteen inches wide. Harvey
says he found plants at Halifax, whose blade was two
to three feet wide. The hold-fast, as in the last species,
is composed of a number of stout roots, put out by
the stem at the bottom. The stem is very slender
and solid at that point, but toward the middle swells
to the diameter of an inch or more, and become
hollow. It tapers also toward the blade to a diameter of half an inch. Altogether, the stem will be
found from six to ten feet long in the full grown plant.
The blade is much the shape and color of the wide
forms of *L. saccarhina*. It grows in deeper water
than that species, and may be found in from five to
ten fathoms or more. It is very abundant from Greenland to Cape Cod, and in the North Pacific.

LAMINARIA FLEXICAULIS, LE JOLIS.

This is the *L. digitata* in part, of Harvey's "Nereis."
The holdfast and stem are much the same as in *L.
saccarhina*, except in the more variable length of the
stem. But the blade is much wider and is split from
top to bottom into several long, strap-shaped segments
from one to three inches wide. The whole blade
may be from one to three feet wide, and from three
to five feet long. It grows in deep tide pools, and
in the sea, from just below low-water mark to considerable depths. This, like the other species of *Laminaria*, puts forth its new, yearly growth in the winter
and early spring, in a most curious way, which I
will now describe.

The new blade grows forth from the top of the
old stem and interposes itself between the old stem
and the old blade. It carries the old blade on its
top, till it has grown to nearly its full size, when
by a process of natural decay, the old blade is separated from the new, and falls away, in the month
of May, and is washed ashore, in great numbers.
The process has a very curious phase in this species.
It is seen that the new frond splits down by a natural process some time before the old blade is cast
off, the old blade, meanwhile, holding the tips of
the straps together at the top, while they are quite

parted asunder lower down. One by one the straps from the margin inwards are pulled away from the old blade, till at last it is held by but two or three central ones. These part at last, and the old frond falls like an autumn leaf

"Because its time to die has come"

Those who live by the sea the year around may be interested to watch this curious process of "shedding the leaf," in this species. It was first described many years ago, by that most pains-taking and sharp-eyed naturalist, Dawson Turner. This species is not common, if it is found at all, south of Cape Cod; north of that it is plenty enough.

LAMINARIA ANDERSONII, EATON.

I have three copies of this plant, sent me a few years ago by Dr. Anderson himself, and for want of a printed description by the author, will give a description of one of these. This specimen is about one yard long. The lower half is a stem with the usual branching hold-fast. The stem is cylindrical, of uniform size, one-sixth of an inch in thickness. It suddenly expands into the blade of the frond which is about an inch wide, and, of course, half a yard long, sides parallel except where it narrows into the stem, broken off or "frayed out" at the top. It is reported at Santa Cruz, California, only, where it grows on rocks with *Pterygophora*.

Genus.— *AGARUM*,* *Bory.*

AGARUM TURNERI, POST. AND RUPR.

This plant differs from the *Laminariæ* among which it grows, by its shorter stem, its thinner blade, its stout midrib running through the whole frond, and, most of all, by the fact that it is perforated throughout with holes of various sizes. This gives it its popular name of "Sea Colander." It grows in deep water, holds to the rocks by a number of root fibres, has a stem one-fourth of an inch in diameter, three to twelve inches long which expands somewhat as it enters the blade, forming a substantial midrib. This blade is usually a foot wide, often more, and from one to three yards long, though you will often find it no more than a foot or half a yard long. It has a rather more pronounced green color than the *Laminariæ*, and, as before remarked, is of thinner substance. It is very abundant

* Agarum = A fungus or mushroom.

SEA MOSSES.

from Cape Cod to Greenland, and is to be looked for among the "Kelp," and other sea-weed thrown up from deep water. It will be known at sight by the frond being full of holes. It is dried and mounted in the same way as the *Laminariæ*.

Sub-Order — *SPOROCHNEÆ.*
Genus — *STILOPHORA,** *Ag.*

There are three species of this genus set down in the books, as belonging to our flora. Only one of them is of sufficient importance to warrant me in making mention of it here.

STILOPHORA RHIZODES,† AG.

Is a plant interesting alike to the botanist and the microscopist; for, if you take its wart-like mass of spores and filaments, and cut a thin section of it, and mount it for the microscope, you will find you have a beautiful object.

It is a filiform plant, with stem and branches once or twice as thick as a bristle. It is much branched by irregular forkings, six or seven times repeated, the extreme ends short and widely spreading. It grows four to six inches high, and is of

---

* Stilophora = Dot-bearing.

† Rhizodes = Root-like.

an olive green color with a yellowish tendency, which is even more pronounced in the dried than in the living plant. Its unmistakable mark is the little wart-like protuberances which are thickly scattered over all the stems and branches, making it decidedly rough to the sense of both sight and touch. It is found on our coast south of Cape Cod only; not very common in most places, but at Orient, L. I., in Peconic Bay, Miss Booth reports it growing in unlimited quantities, in July and August.

Sub-Order.— *ASPEROCOCCEÆ.*
Genus.— *ASPEROCOCCUS,*[*] *Lam.*

There are two species of this genus on our eastern coast and one in California. Only one is common with us here; the other, therefore, *A. compressus,* which has been reported only at Gloucester, will not be described.

Asperococcus echinatus,[†] Grev.

Frond flat or inflated, from three inches to one or two feet long, and from one-eighth to half an inch wide; blunt at the apex, and attenuated toward the base. It may be known by its light olive color

* Asperococcus = Rough-seeded.
† Echinatus = Prickly.

and by being covered all over on both sides with minute, oblong dots of a darker shade, which are masses of spores. This roughening of the surface by these spore masses, gives the plant both its generic and specific name. It is a summer annual and grows on the rocks, in pools between tides. Mr. Collins has collected it at Revere and Nantasket, from June to August; Mrs. Davis, at Gloucester in the spring. I have found it in the summer at Marblehead, but not very common.

ASPEROCOCCUS SINUOSUS, BORY.

This plant much resembles our *Leathesia tuber-formis* in outline and habit of growth, though it is much thinner in substance, and grows in much larger clusters. Harvey says each individual frond is globose, one or two inches in diameter or larger, becoming much inflated and irregular in outline as it advances in age, and is thus often ruptured and pierced here and there with holes of irregular shape and size. The frond is membranous, thin, soft, but not very tender; color, a brownish olive. It may be found common all along the California coast, at all seasons, growing, Dr. Anderson says, on tips of *Halidrys*. Mrs. Bingham finds it growing on small rocks and other Algæ at mid-tide. Dr. Dinnick on

Amphiroa. Mr. Cleveland, in bunches, on flat rocks between tides, and washed ashore on the beach.

Sub-order.— *CHORDARIEÆ.*

Genus.— *CHORDA,** *Lam.*

CHORDA FILUM, STACK.

The thread-like cords, which are sometimes popularly called "Dead men's lines," and sometimes "Mermaids' fish-lines," are plants very easily described and very easily recognized. The frond of *C. filum* is a single undivided cord rising from a discoid hold-fast, by which it is attached to some small pebble or shell upon the sea bottom. At first, a mere thread, it increases in size till it is as large as a pipe-stem, or larger, then again tapers to a long, slender-pointed termination. When young, it is covered all about with short, fine, olive-colored hairs, which disappear in age. It loves quiet waters and grows to the height of ten, twenty, and even forty feet, according to favorable conditions. It is quite tough and somewhat elastic when recent. It is a favorite habitat of some of the smaller Algæ, like some species of the *Ectocarpus,*

* Chorda = A cord.

Callithamnion etc. The Cyclopœdia Britannica mentions the fact that it is distributed in beds through the North Sea and British Channel, fifteen to twenty miles long, and yet not more than 600 feet wide. It is common along all our shores, from New York northward. It grows, of course, in deep water. Its fronds reach up, at least, to the surface. The old fronds should be allowed to dry off a little before mounting, but the young ones, covered with hairs, may be floated out in water. The long plants are best disposed of by coiling up neatly on the sheet of mounting paper, and drying in the usual way, under pressure. They seem to adhere well.

Genus.— *CHORDARIA,* * Ag.

CHORDARIA FLAGELLIFORMIS, AG.

The *whiplash Chordaria* is found in bewildering abundance along our whole coast. It may be known by its very dark brown or quite black color, both in the water and on paper ; and by its long, slender, naked, mostly undivided branches, which sweep off from all sides, and, in not ungraceful curves, over-

---

* Chordaria = Cord-like.

arch the top of the frond. Neither stem nor
branches are ever larger than a pack-thread, and
commonly not half so large. The leading stem
ascends half-way or more, through the whole length
of the plant. The branches put out very irregularly
all around; sometimes scattered, sometimes much
crowded, sometimes short, but more often long and
bent inward, as indicated above. It grows upon
shells, stones, rocks and other Algæ, to which it is
fastened by a minute disk. The substance of the
frond is cartilaginous, tough and elastic. When
taken from the water it will be decidedly slippery
to the touch, and when carried home and removed
from the mass of plants in the collecting case, it
will be found to be not a little slimy. It will be
quite sure to stain the cloth used in pressing and
drying it, and, perhaps, also the paper on which it
is mounted, a dark, brownish color. It is an annual,
and grows between tides, not usually over a foot
high, and the old fronds will be quite certain to be
infested with some species of *Ectocarpus.*

CHORDARIA DIVARICATA, AG.

The widely branched *Chordaria* is a deep-water
plant and may be collected along our whole coast,
from New York to Gloucester, and probably farther

north. But it will be found more plentiful south
than north of Cape Cod. I have taken it at
Southold, L. I., and at Wood's Holl. It is not so
robust a plant as the last. From the first, it branches
out widely in all directions, in a straddling, strug-
gling, bushy way. The branches, which branch again
and again, are beset throughout with short (one-
sixteenth to one-tenth of an inch), spines, which
are mostly forked widely at the ends. These are
the characteristic points. The plants of this, like
those of the last species, are somewhat slippery and
slimy, and must not be put under too much pressure at first. It often grows a foot or more, though
my specimens are not more than half that height.
My correspondents report it as found all summer at
all points.

CHORDARIA ABIETINA, RUPR.

This is the only species of this genus found on the
coast of California. It is quite common at Santa
Cruz and northward, growing on the boulders, along
rocky beaches.

A mounted specimen, four inches high, lies before
me as I write. It has a principal leading stem extending the whole length of the plant, which is two or
three times as thick as a bristle, and much attenuated

/

at the base. A quarter of the way up it is bare. From
that point it is thickly beset all around with short
branches, varying from half an inch to one and one-
half inches long, undivided, narrowly constricted at
the base, blunt at the apex, mostly curved, and stand
out perpendicularly from the main stem.

Genus.— *CASTAGNEA, Thuret.*

CASTAGNEA ZOSTERÆ, THURET.

This species is named from the "Eel grass" or
Zostera, on the fronds of which it commonly grows.
It is a very slender plant, not larger than a thread or
bristle, and some six or eight inches long, of a light
olive color, somewhat bent in a zigzag way, and but
sparingly branched. The branches are irregularly
placed, short (about one inch long), spreading horizon-
tally from the main stem, and either widely forking
or beset with twig-like branchlets, which are also frequently forked or spiney. It adheres nicely to paper,
and is not an uninteresting though by no means a
handsome plant. I found it in August, in Marblehead
harbor. My correspondents do not report it else-
where, though Dr. Farlow records it in Wood's Holl,

and Mr. Collins and Mrs. Bray in Robinson's "List of Essex Plants," report it from Gloucester.

CASTAGNEA VIRESCENS, THURET.

This is apparently a shorter but more robust plant, and more thickly branched than the last.   It is of a more pronounced green color, as its name implies. It is not more than three inches long, main stem and branches both straighter than in *C. Zostera*, but having the twiggy appearance peculiar to the genus. American plants are said to grow on *Zostera*, though no doubt it grows parasitical on the other Algæ also. According to Le Jolis they are found, on stones and pebbles, and in tide pools on the rocks at half tide, toward the end of spring.   Mrs. Davis finds it growing on sand covered rocks at half tide, all summer at Gloucester, and Mr. Collins found it in June at Revere, cast up from deep water, not very common.   Miss Booth makes report of it in the same situations at Peconic Bay.   It is also reported at Wood's Holl and Portland.   I should expect to find it at Marblehead.

Fig. 1.

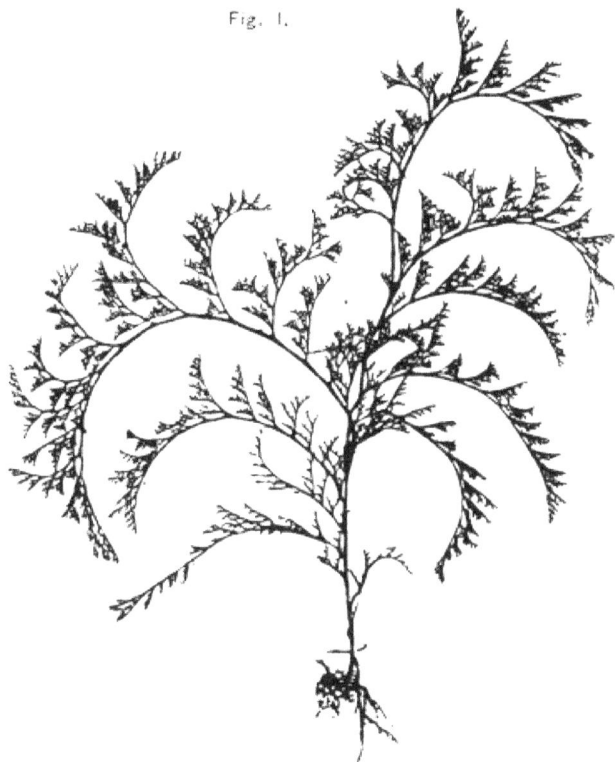

Fig. 2

1. POLYSIPHONIA PARASITICA, *Grev.*
2. MICROCLADIA BOREALIS, *Rupr.*

Sub-order.— *MYRIONEMÆ.*
Genus.— *LEATHESIA*, *Gray.*

LEATHESIA TUBERFORMIS, GRAY.

I suppose it was thought a great compliment to
a brother naturalist, to name this plant for him.
But one cannot help thinking, that one would
rather lend his name to some of the more
interesting and beautiful of the "flowers of the
sea." Still, this plant has beauties of no uncommon
kind, as you would see, if you were to take a very
thin slice of it, and put it under the lenses of a microscope. It is also very widely distributed, being
found in almost every sea, and on the most distant
shores of the whole globe. So this humble and homely
plant, carries the name of the Reverend Naturalist,
G. R. Leathe, far and wide. To the unaided eye, it
looks as it lies fastened there upon the rocks, or
resting its green lobes upon the fronds of *Chondrus
crispus*, so nearly like an unripe tomato, that you
are inclined to doubt if it can be an Alga at all,
and are more than half disposed to believe, that it
must be some succulent vegetable which Neptune is
preparing for his board. It makes its appearance in
April or May, and is ripe by August or September,
and then soon disappears.

Genus.— *ELACHISTA,** Duby.

ELACHISTA FUCICOLA, FR.

No doubt you will wonder what the little tufts of olive colored hairs are, which are so common upon the " Rockweed," every hair of which seems to radiate unbranched, from some central point of attachment hidden in the tuft. I have given its name above. It will be noticed also that, though the longest hairs are not over half an inch long, there is a mass of them much shorter than that, above the general crop of which, the long ones seem to stand out stiff and solitary. It had better, perhaps, be removed from the *Fucus* before mounting, though a thin slice of that might be cut off with the *Elachista.* It makes a very interesting microscopical object. Its delicate pencils may be found upon the " Rockweed " almost everywhere, for it is widely distributed.

Sub-order.— *SPHACELARIEÆ.*
Genus.— *CLADOSTEPHUS,*† *Ag.*

CLADOSTEPHUS VERTICILLATUS, AG.

The *whorled Cladostephus* is very easily distinguished from all other plants of the sea, except

---

* Elachista = The smallest.
† Cladostephus = Branch crowned.

its "next of kin," the *C. spongiosus ;* and it is not
of the first importance, if it is not distinguished from
that, for it is doubtful if they are quite distinct species.
The frond is not much thicker than a bristle, quite
cylindrical, hard and stiff. It begins to branch quite
low down, and continues, by repeated, regular, though
not wide forkings. The whole frond is clothed though-
out with a fleece of densely set, very short branchlets,
which grow in regular circles around the plant. The
circles or "whorls" are not more than one-tenth of
an inch apart, and the branchlets are not less
than one-eighth of an inch long, somewhat incurved,
hugging the stem closely about, and those of one
"whorl" overlapping the bottom of the row next above
it. This gives the whole plant a decidedly spongy
quality to the sense of both sight and touch. It grows
on the rocks, nearly down to low-water mark. Color,
brownish olive. Height, three to five inches. It is
a perennial and fruits in winter. I found it and *C.
spongiosus*, growing together in great abundance, on
the low rocks, east of the first beach at Newport. I
also got several fine specimens of it at Martha's
Vineyard. It is said to belong to our whole New
England coast; but I think it must be rare in our
northern waters, for I have collected Algæ along the
shores of Salem, Marblehead and Nahant, several

years, and have never found it growing there. None of my correspondents have reported it north of Cape Cod.

CLADOSTEPHUS SPONGIOSUS, AG.

This plant differs from the last by its shorter habit; by being more irregularly branched, the branches spreading more widely, and having a thick, clumsy, rambling appearance, and by the branchlets being longer, irregularly whorled, and clothing the frond in a denser, spongier fleece. It is not at all unlikely that intermediate forms might be found which should connect the extremes, typical of these two species, in a single graduated series. My European plants appear decidedly more "spongy" than the American. Its local habitat is the same as that of *C. verticillatus.*

Sub-Order.— *ECTOCARPEÆ.*
Genus.— *ECTOCARPUS,** *Lyngb.*

According to Dr. Farlow's list, this genus, in our American waters, includes fifteen species. Of those I have selected five of the most common for our study. These plants, like the *Cladophoræ* in the green Algæ, and the *Callithamnia* in the red, are of capillary or hair-like fineness, and like them are

* Ectocarpus = External fruits.

composed of cells put end to end in a single series.
The determination of species is made, in most cases,
by the appearance of the fruit masses, (*propagula*),
and by the peculiarities of the branching. These
points can best be determined by the use of the
compound microscope, but they can be made out
with a good pocket lens. They are mostly parasitical
on other Algæ, *Fucus*, *Chorda*, *Chordaria* and
Zostera, etc. The color of the smaller forms is very
apt to be a fine olive green.

ECTOCARPUS FIRMUS, AG. (*E. littoralis, Harv.*)

This is said to be the commonest species of the
genus on our coast, and grows parasitical on the
littoral *Fuci*. The tufts are of various lengths up to
ten or twelve inches, dense, filaments fine, interwoven,
much and irregularly branched; branches mostly
alternate, repeatedly divided, the divisions made at
acute angles, the upper ones opposite; articulations
of branches almost as long as broad. The *propagula*
form elongated linear swellings in the substance of
the greater and lesser branches, many times longer
than broad. Color varies from olive green to brown.
Found at all seasons.

8

ECTOCARPUS FARLOWII, THURET.

This is a shorter and somewhat coarser plant than the preceding, growing in the same situations upon *Fucus nodosus.* In my specimens, the end of the *Fucus* is clothed, for the space of three inches or more, with a dense, dark green mass of *Ectocarpus* filaments, half an inch long. I have seen no detailed description of the plant; but perhaps its outward appearance, as given above, being somewhat distinct and well-marked, would serve most collectors as a clue to identification, better than a fuller account of the fruit and branching. I found it common at Marblehead, in the summer. It is also found along the coast north, as far as Peak's Island, Maine.

ECTOCARPUS SILICULOSUS, LYNGB.

This plant is very common along our whole eastern coast, and is found occasionally on the Pacific shores. It grows on various substances between tides, but seems especially to affect the string-like fronds of the *Chordaria flagelliformis.* The color is mostly a yellowish green, but variable. Fronds from three to six inches long, not entangled, filaments very slender, and excessively branched, all the divisions alternate with acute axils. The *propagula* are formed by the transforming of a portion of the ultimate ramuli, that portion commonly nearest the end, into spore masses, which, under the glass, look not unlike minute ears of corn.

ECTOCARPUS VIRIDIS, HARV.

This may be a mere variety of the last. It grows in the same situation, but is much less common. The color is a more pronounced green, and the frond is decidedly more feathery, loose, open, and expanding, than in *E. siliculosus*. The *propagula* are the same, only that they are formed in the base of the ultimate ramuli and so have the unchanged portion extending beyond the spore mass. Our figure in Plate IV., gives a very good representation of this beautiful species.

ECTOCARPUS TOMENTOSUS, LYNGB.

This is a native of our northern waters. The filaments are fine, twisted and matted together like cords, or interwoven into a dense sponge-like branching tuft. Articulations two or three times as long as broad. *Propagula*, oblong, obtuse set on the lower branches by a short stem. Color, from yellowish olive to dark brown. It grows on various substances between tides. It may be looked for throughout the season.

Sub-order.— *DICTYOSIPHONIEÆ.*

Genus — *DICTYOSIPHON,** *Grev*

DICTYOSIPHON FŒNICULACEUS, GREV.

This is our only species of this genus. It grows
in rock pools and below tide, and occurs from L. I.
Sound northward, but is more common in our
northern waters.   Frond filiform, about as thick as
a bristle ; harsh to the touch ; from six inches to
two feet long ; profusely and irregularly branched
on all sides from top to bottom.   The primary
branches are long, and closely beset with secondary
branches which are also long and straight, and often
of hair-like tenuity.   Color, a brownish olive, dark
when dry.   It adheres pretty well to paper in drying.   Mr. Collins collected it from March to September, at Nahant and Nantasket.   I found it not
uncommon at Marblehead, all summer, and Miss
Booth reports it in Peconic Bay, **L.** I.   Others
have found it at Boston and Newport.   It certainly
may be expected in favorable localities all along the
coast.   It is not noted for its beauty as a herbarium specimen.

* Dictyosiphon = A netted tube.

Sub-order — *DESMARESTIEÆ.*

Genus.— *DESMARESTIA,** *Lam.*

Of this genus we have four species, divided
equally between the two oceans. The cylindrical
and narrow forms belong to the Atlantic and the
flattened or strap-like forms are natives of the
Pacific. It is not a little singular that one species,
D. ligulata, should be very common on the eastern
shores of both the Atlantic and Pacific oceans, and
not found at all on the coast lying between, viz.,
the western shores of the Atlantic.

DESMARESTIA VIRIDIS, LAM.

This is a large and fine plant, growing from one
to three feet in hight, of a beautiful chestnut olive
color when fresh, turning to a dark green when
dry. It is found on rocks, stones, and other Algæ,
in tide pools near low water mark, and in deep
water. The frond is cylindrical or filiform, twice as
thick as a bristle in a plant two feet long, beset, at
rather remote intervals, with long, primary branches,
which come out in pairs exactly opposite each other
on the two sides of the main stem. These branches
are themselves branched in the same way by pairs

* Desmarestia was named for Desmarest, a French Naturalist.

of opposite secondary branches, and these again in like manner by their branchlets. All the divisions are long and the ultimate parts very fine and hair-like. Indeed, a large and beautiful plant in my herbarium presents an appearance not unlike that of long, wavy tresses of hair. If it never received the popular name of "Mermaid's hair," it is quite time it was christened that. It is reported very common along all our northern shores, from February to November, and less common in southern waters in the summer.

DESMARESTIA ACULEATA, LAM.

This plant is found the year around, growing at low tide and in deep water. It is very common so that special localities need not be named. Frond, cylindrical at base, but soon flattening; in a plant a foot and a half high, as thick as a sparrow's quill. Branches, alternate, irregular, half forking, much flattened, from one-twelfth to one-eighth of an inch wide two or three times sub-divided. The young plants, and apparently the younger parts of all the plants, are clothed with opposite pencils of fine, beautiful olive-green filaments, from one-sixteenth to one-half an inch long. A larger plant before me, collected at Marblehead, Mass., in August, has them very short;

and a smaller plant from the island of Spitzbergen, collected July 23rd, has them half an inch or more long.· When these pencils fall away, they are replaced by short, sharp, awl-like spines, set regularly and alternately on each edge of the flattened branch, pointing forward. It is, perhaps, an arctic plant, but it is found in temperate waters, south of Cape Cod. It is said sometimes to attain a height of six feet. It is an interesting plant, and the young forms are very beautiful, and adhere nicely to paper in mounting.

DESMARESTIA LIGULATA, LAM.

This is the most common California species, and exceeds in interest, if not in beauty, either of our Atlantic plants already named. It grows a foot or two high, flat, one-fourth to one-half inch wide, beset, at intervals, along the edges, by pairs of opposite flat branches. And these, again, are more thickly clothed by shorter, flat branchlets, serrated along the edges with sharp, forward-pointing teeth.

Both the primary and secondary branches are narrowed to a point at base and apex. The substance of the frond is thin and delicate ; the color, a yellowish olive, in the specimens which I have

seen. It grows in great abundance, at low tide and
below, on rocks, along the whole California coast.
Mr. Cleveland says it is washed up from deep
water, and lies in great heaps on the beach, near
the Mexican boundary of Southern California.

DESMARESTIA LATIFRONS,* KUTZ.

This plant seems to occupy a middle ground between *D. aculeata* and *D. ligulata,* having branches
shorter and wider and less numerous than the former,
and much narrower and thicker than the latter.
The branching is alternate, like that of *D. aculeata,*
and the secondary branches have the same remote
alternate sharp spines of that species. In the frag-
ment of a plant before me, which is about six
inches long, the stem is one-tenth of an inch wide,
primary and secondary branches about the same.
Both main stem and primary branches appear under
the lens to be "midribed." It is not a very rare
plant at Santa Cruz and in the north of California,
but grows at low-tide mark, on the rocks, at all
seasons. At Santa Barbara it is very rare, and has
not yet been found at San Diego.

* Latifrons = A wide frond.

POLYSIPHONIA BAILEYI, *Ag.*

Sub-order — *PUNCTARIEÆ.*

Genus.— *PUNCTARIA,* * *Grev.*

PUNCTARIA LATIFOLIA,† GREV.

Fronds, pale olive green; thickish, membraneous, soft and tender, more or less dotted with minute spore masses, suddenly tapering at the bottom, from one to three inches wide in the broadest point, and from eight to twelve inches long, the proportions the same in the smaller plants. When young, the substance is thin and soft, and almost gelatinous to the touch, being then covered with very short pellucid, almost invisible hairs. In that state it is of a light olive green color. When older, it gets darker. The margin of the frond wavy, and in old plants the substance of the frond is thicker and more rigid. In that condition it will be distinguished from plants of the next species chiefly by its sudden narrowing at the base.

It is a summer annual, growing between tides on stones and Algæ. It will be met with most commonly in the var. *Zosteræ,* or *P. tenuissima,* of Harvey's "Nereis," a small form, not more than two or

* Punctaria = Dotted.
† Latifolia = Wide-leaf.

three inches long and one-fourth of an inch wide,
very thin and delicate, fringing both edges of a
blade of *Zostera*, or growing in the same manner
from the sides of a frond of *Chorda filum*. Mr.
Collins finds it in deep water and on *Zostera*, at
Revere, from April to July; Mrs. Davis, from
April to November, in rock pools everywhere about
Gloucester. I have a copy of the typical form
collected by Mr. A. R. Young, at College Point,
L. I., in May. It was collected by Mr. Hooper, at
Fort Hamilton, New York Bay, and at Flushing
Bay, by Prof. Bailey.

PUNCTARIA PLANTAGINEA,* GREV.

Frond, dark brown, leathery, much attenuated at
base from near the middle, blunt or wedged-shaped
at the top, from six to twelve inches long and
from one to one and a half inches wide. It is a
summer annual, and grows on stones and other Algæ,
between tide marks and below. It is not so common as the last, but I have it reported all along
our north eastern seaboard.

It does not usually adhere well to paper, and it
is far from being an inviting specimen to person
whose interest in these plants is other than scientific.

* Plantaginea = Like the Plantain.

Sub-order.— *SCYTOSIPHONEÆ.*

Genus.— *PHYLLITIS** (*Kutz.*), *Le Jolis.*

PHYLLITIS FASCIA,† KUTZ.

This is quite common along our rocky shores, at all seasons, in tide pools near low-water mark. It usually grows in tufts: a cylindrical stem gradually expands into a long, flat, narrow frond, from one-fourth to one inch wide, and from three to twelve inches long. It is usually blunt at top, and, as just said, attenuated below. My specimens are narrow, with parallel sides, one-third of an inch wide and twelve inches long. The color is a brownish olive, and the substance membraneous, but not very thick. My Californian correspondents report it very common along the whole extent of that coast.

Genus.— *SCYTOSIPHON,‡ Lyngb.*

SCYTOSIPHON LOMENTARIUS, AG.

This species grows in much the same situations as the last, oftentimes in company with it, in the tide pools. It is common on our eastern coast, and is

---

* Phyllitis = Leaf, like Hart's tongue.

† Fascia = A band.

‡ Scytosiphon = A leather tube.

reported the same in California. It grows from
eight to eighteen inches high, cylindrical, unbranched,
attenuated at top and bottom, one-fourth of an inch
in diameter, inflated, and sharply and definitely constricted at irregular intervals, which gives it the
appearance when growing, of a string of small, narrow
bags tied together by the ends. Color, a brownish
or greenish olive. Substance, membraneous and soft.

      *   *   *   *

There are no more fitting words with which to
bid adieu to this modest-hued, homely, often coarse,
but always interesting group of plants, than these of
the Poet, who loves the sea and the

SEA WEED.

"When descends on the Atlantic
    The gigantic
Storm wind of the Equinox,
Landward in his wrath he scourges
    The toiling surges,
Laden with sea weed from the rocks.
Ever drifting, drifting, drifting,
    On the shifting,
Currents of the restless main;
'Till in sheltered coves, and reaches
    Of sandy beaches,
All have found repose again."

                    Longfellow.

KEY TO THE
GENERA OF THE ATLANTIC COAST.

RED ALGÆ.

I. FROND MEMBRANEOUS.

1. *Frond Midribed.*

(*a.*) Plants small, with regular veins from midrib to margin of frond. *Delesseria.*

(*b.*) Plants large, without veins, midrib slender. Frond thin, brilliant pink, more or less sprinkled with darker colored dots.

Grinnellia.

2. *Frond Stalked.*

Membrane small, short, forked, growing on the apex of branching, cylindrical stems. *Phyllophora.*

3. *Frond plain, Membrane smooth, without stalk, midrib, or vein.*

(*a.*) Frond large, thickish, mostly wedge or fan shaped, palmately divided, sometimes strap-shaped. " Dulse."

Rhodymenia.

(*b.*) Frond thin, tapering to top and bottom, bearing on the edges toothed frondlets of the same shape. *Calliblepharis.*

II. Frond flattened or compressed.

1. *Frond forked.*

(*a.*) Small, short, wedge-shaped, once or twice forked.

(i.) Frond thick, smooth, purple or green.

"Irish Moss," *Chondrus.*

(ii.) Frond channeled, more or less covered with *papillæ*, dark.   *Gigartina.*

(iii.) Frond stalked, thin, narrow, red.

Gymnogongrus.

(*b.*) Frond long, narrow, partly cylindrical, many times divided.   *Gracilaria.*

2. *Frond pinnately divided.*

Plant small, pinnæ and pinnulæ, fine and set in one plane.   *Ptilota.*

3. *Frond irregularly divided.*

Frond forking and branching irregularly, profusely, mostly in one plane, from a marginal point.   *Euthora.*

III Frond filiform or thread-like.

(From size of sewing cotton to that of wrapping twine, branched).

1. *Plants whose ultimate branchlets taper to both ends.*

(*a.*) Plants with one main or leading stem.

(i.) Main stem mostly undivided, bare at base, clothed above with simple unbranched ramuli. *Halosaccion.*

(ii.) Robust, coarse, profusely branched, branches often ending in twining tendrils, dull brown or purple, very common ; six to ten inches high. *Cystoclonium.*

(iii.) Smaller, finer, branches shorter, full red or pink, rare. *Gloiosiphonia.*

(*b*.) Plants without leading stem.

(i.) Large, smooth, robust, two or three times divided ; ramuli long, slender at point, slightly curved ; reddish purple to pink ; prominent fruit vessels in ramuli. Plant six to twelve inches high. *Rhabdonia.*

(ii.) Small, slender ; ramuli long, curved ; beautiful delicate pink. Plants three inches high. *Lomentaria.*

(iii.) Larger, brownish, slender or robust ; branches long, ramuli very short, often minute. *Chondriopsis.*

(iv.) Slender, brown, branches long, bare and hooked at the ends ; ramuli short. *Hypnea.*

2. *Frond regularly forking.*

(*a.*) Long, elastic, worm-like, axils wide and rounded. *Nemalion.*

(*b.*) Short, stiff, black, widely forking, uniform size, not adhering to paper. Three or four inches high. *Polyides.*

(*c.*) Same outline, soft, adheres, rosy red. *Scinaia.*

3. *Plants clothed with fine hairs.*

(*a.*) Stem robust; branches few, long and mostly simple. All parts thickly clothed with brilliantly colored pink or purple fine hair, like "Chenille." *Dasya.*

(*b.*) Stem and branches slender, several times divided; hairs much paler, shorter and less abundant. *Spyridia.*

4. *Fronds many times and finely divided, robust or slender, mostly dark or brown.*

(*a.*) Ultimate ramuli, often in clumps or minute brushes, black or brown. *Rhodomela.*

(*b.*) Plants variously, but profusely branched, mostly fine, often arborescent, fruit vessels pear-shaped; black, reddish or light brown. *Polysiphonia.*

5. *Frond consisting of visibly articulated, or jointed filaments.*

> Slender or robust, branching or forking; filaments showing alternately white and red, or light and dark bands.
>
> > *Ceramium.*

6. *Frond stiff, wiry, black.*

> Intricately and irregularly branched, sometimes bleached white. *Ahnfeltia.*

7. *Frond stony and hard.*

> Purple to white. *Corallina.*

IV. FROND CAPILLARY.

> (Composed of a single series of cells placed end to end).

1. *Cells long.*

> Frond divided by regular, narrow forkings, fan-shaped, level topped; color pale, delicate pink. *Griffithsia.*

2. *Cells short.*

> Plants mostly small, often shaped like a miniature shrub; much branched, final divisions as fine as cobweb; color brilliant red or pink, the most beautiful of plants. *Callithamnion.*

KEY TO THE GENERA OF THE PACIFIC COAST.*

I. Frond Membraneous.

1. *Frond plain, mostly undivided, smooth, or roughened only by seed vessels.*

(*a.*) Thick, large, reddish brown.
Sarcophyllis.

(*b.*) Thinner, large, purplish color. *Iridæa.*

(*c.*) Undivided, branched or cleft; brown, purple, or green. *Grateloupia.*

2. *Frond thick, covered with pappili.*
Undivided, forked or irregularly branched, deep red, or purple. *Gigartina.*

3. *Frond narrower, thick, leathery, smooth.*
Sword-shaped leaflets from side or end of main frond; dark red brown.
Prionitis Andersonii.

4. *Frond much divided.*

(*a.*) Thin, deeply lobed, or forked, mostly dark red; not adhering well.
Nitophyllum.

(*b.*) Thicker, more intricately divided, more brilliant red color, adheres.
Callophyllis.

Only those Genera which have species peculiar to the Pacific Coast are included in this Key, the rest will be found in the other.

ξ. *Fronds regularly forking, thin, narrow; sides of lobes parallel, ends rounded.*

(*a*.) Dull red, not adhering. *Rhodymenia.*

(*b*.) Brilliant red; interrupted midrib of darker color, or fruit dots scattered over the surface; adheres. *Stenogramma.*

II. FROND F ATTENED OR COMPRESSED.

1. *Frond pinnately branched.*

(*a*.) Frond narrow, dense, hard, dark red. Primary branches, alternate or forking; secondary, short, tapering to both ends, pinnate. *Prionitis lanceolata.*

(*b*.) Frond narrow, cartilaginous, divided into several branches; pinnæ and pinnulæ, alternate, blunt at apex; dull purple. *Laurencia.*

(*c*.) Pinnæ, arranged on the edges of the main stem and long branches, short, the opposite ones unlike. *Ptilota.*

(*d*.) Frond very narrow, horny when dry; main branches irregular; pinnæ and pinnulæ exactly opposite, with wide rounded axils, ultimate pinnæ tapering to both ends; purple, often faded. *Gelidium.*

2. *Fronds irregularly branched.*

(*a*.)   Frond leathery, narrow, very dark reddish
brown; branches in one plane, flat,
narrowed at base and top, bent sword-
shape, and often bordered with fine
spines; eight to twelve inches high.

Farlowia.

(*b*.)   Plants smaller and narrower, branching
much the same as the last; secondary
branches, bordered with incurved spine-
like ramuli, much attenuated at both
ends. Color, very dark red.   *Pikea.*

3. *Frond with leading stem.*

Branches long, alternate; secondary, short,
alternate; ultimate ramuli, alternate, incurved, awl-shaped, not constricted at
base.   *Microcladia.*

III.   FROND FILIFORM OR CYLINDRICAL.

1. *Frond coarse, thick as pack thread.*

(*a*.)   Frond divided by regular forkings, several
times repeated; horny when dry, dark.

Ahnfeltia.

(*b*.)   Frond with leading stem, branches short,
stout, tapering at both ends. Clear red.

Rhabdonia.

(*c.*) Stem branched and forked; end of
branches beset with many short, stout,
oval or obtuse ramuli. *Chylocladia.*

2. *Frond finer and more elaborately divided.*

(*a.*) Stem robust, branches irregular; ultimate
ramuli, clustered in bunches; black.

Rhodomela.

(*b.*) Frond delicate, many times finely and
pinnately divided; color, brown or black.

Polysiphonia.

(*c.*) Frond delicate, finely pinnated, brilliant
pink *Callithamnion.*

The night is calm and cloudless,
   And still as still can be,
And the stars come forth to listen
   To the music of the sea.
They gather, and gather, and gather,
   Until they crowd the sky,
And listen in breathless silence,
   To the solemn litany.
It begins in rocky caverns,
   As a voice that chants alone
To the pedals of the organ
   In monotonous undertone;
And anon from shelving beaches,
   And shallow sands beyond
In snow-white robes uprising,
   The ghostly choirs respond.
And sadly and unceasing
   The mournful voice sings on,
And the snow-white choirs still answer,
      Christe Eleison!

                          Longfellow.

CHAPTER IV.

———:0:———

RED ALGÆ.

.

CHAPTER IV.

Sub-class.— *RHODOSPORÆ* or *FLORIDEÆ*.

WE have now come to the Red "Sea Mosses." They are more highly organized than the plants we have been considering. This is apparent in the greater variety of form, and complexity of structure, as well as in the higher and more elaborate machinery for the reproduction process, which is seen in them.

The Red "Sea Mosses" are characterized by the presence of two different kinds of seeds, or spores. One kind is produced by a process analogous to that by which seeds and fruit are produced in the flowering plants; that is, by the presence and co-operation of a staminate and pistillate element. This is the

sexual fruit, and usually appears in minute clusters
upon the branches of fertile fronds, or else encased
in little egg-shaped baskets, or other receptacles. It
is also not unfrequently found embedded in the substance of membraneous fronds, or held in wart-like
protuberances which arise from their surface.

The other or asexual spores are produced, apparently, by a change in some of the vegetable cells
of the plant. They always appear in groups of four,
hence their name, " *Tetraspores* " or " *Tetragonidia.*"
The original, or " Mother cell," seems to part its
contents invariably into four secondary cells, and each
of these is capable of reproducing the plant. They
are found in various situations, but, except in some of
the lower plants of the group, always occur embedded
in the substance of the frond. It is a rule, which so
far as I know, has no exception, that the two kinds
of fruit never appear upon the same individual plant.

The Red Mosses will no doubt make up the principal part of all your collections. Certainly they are,
as a general thing, more interesting and more beautiful,
and appear in much greater variety of form, than those
of the other classes. Some of them are marvelously
fine and delicate, and make the most exquisite and
fairy-like pictures when spread out upon paper. The
wonder is, how such fragile things can find the means

and opportunity to live and grow in the rough, tumul-
tuous and stormy sea. But you will not long have
been an observer of the ways of Old Ocean without
often seeing what the Poet has so finely told in the
following lines :

SEA TANGLE.

"Go show to earth your power!" the East Wind cried
Commanding; and the swift submissive seas,
In ordered files, like liquid mountains, glide,
Moving from sky to sky with godlike ease.

Below a cliff, where mused a little maid,
It struck. Its voice in thunder cried "Beware!"
But, to delight her, instantly displayed
A fount of showering diamonds in the air.

*   *   *   *   *   *   The wave passed on;
Touching each shore with silver-sandled feet,
But tossed, in flying, in the sun which shone,
A handful, to her lap, of sea-blooms sweet.

More delicate than forms that frost doth weave
On window panes, are Ocean's filmy brood:
Remembering the awful home they leave,
Their hues to that dim underworld subdued.

Fair spread on pages white, I saw arrayed
These fairy children of a sire so stern;
Their beauty charmed me; while the little maid,
Spoke of her new found love with cheeks which burn.

"So grand, so terrible, how could I know
He cared for these?" she faltered,— "darlings dear!
That his great heart could nurture them and glow
With such a love beneath such looks severe?"

Like God, the Ocean, too, the least can heed,
Yearn in a moon-led quest to farthest shores,
And fondle in delight its smallest weed,
Yet look to Him it mirrors and adores.

J. G. Appleton.

Order.—*RHODOMELEÆ.*

Genus.— *DASYA,* Ag.*

DASYA ELEGANS,† AG.

Of this genus but one species is found on our Atlantic coast, within the geographical limits which this book is intended to cover. But, happily, this is the most interesting and beautiful representative of the genus, known to our American flora, viz., the *Dasya elegans.*

It is sometimes popularly called "chenille," because in the water it looks very like a piece of that sort of finery. No one acquainted with the appearance of chenille, would, for an instant, mistake a specimen of this *elegant Dasya*, when seen floating in its native element. Out of the water, lodged wet upon the rocks, or mixed with other Algæ, it looks more like a stringy mass of pink or purple

* Dasya = Hairy.
† Elegans = Elegant.

jelly. The artist has made an excellent representation of a beautiful specimen of this plant, in our Plate V.

The body of the plant is a robust, sparingly but irregularly branched cord, from six inches to two or three feet long, and from once to three times the thickness of a pack-thread. The branches are long, and mostly undivided, and the whole plant is clothed with a fine, delicate body of purple-lake colored hairs, from an eighth to a third of an inch in length. This gives it the appearance of chenille. When a little faded, this fine, silky plush assumes a delicate or bright pink color. The plant grows attached, by a discoid hold-fast, to rocks, stones, wood-work, and other Algæ, from low-tide mark to a depth of several fathoms. It is not found north of Cape Cod, but may be looked for in all waters south of that point. I have collected it, in July, at Fort Hamilton, and along the beach toward Coney island, in great abundance — splendid fronds, two feet long — along with that most brilliant American Alga *Grinnellia Americana.* I have collected it also in fine condition at Newport, east of the first beach, as late as October 4th. In a breezy but not unpleasant walk, which I took along the shore from Falmouth to Wood's Holl, beneath a

gray, November sky, and the sea a steel blue, cold
and angry, I found this among the most plentiful
of the late autumnal "Sea Mosses." Displayed with
taste, it makes an elegant picture on paper. A
comparatively light pressure should be put on it at
first, in drying, else its tender frond will be crushed
and ruined.

Genus.— *POLYSIPHONIA,* * *Grev.*

This is the largest genus of Red Algæ. Agardh
in his latest work, enumerates no less than 129 un-
doubted species. Many more have been proposed
by other writers. About thirty species belong to
our American flora. But several of them are peculiar to the sub-tropical region of Florida, and will
not come within our reach. Others are too rare or
insignificant to be enumerated in this work. But
all such as are likely to be met with, at all
common, will be described. The color of these
plants ranges between the browns and a full black ;
only three, herein described, show traces of red :
P. urceolata, commonly, and *P. violacea* and *P.
Olneyi,* occasionally. On the fertile fronds, the
beautiful, little egg-shaped fruit-holders will be easily

* Polysiphonia = Many tubes; referring to the internal structure of the
frond.

discovered with the naked eye. The *Polysiphoniæ* form a marked feature of the marine flora of every sea.

POLYSIPHONIA FASTIGIATA, GREV.

The *pointed Polysiphonia* is very common on the north Atlantic coast, growing as a parasite on *Fucus nodosus*, and rarely on *F. vesiculosus*. Prof. Kjellman reports it growing on *Halosaccion ramentaceum*, in Spitzbergen. It looks not unlike a little dark brown or black ball or tassel, attached to the ends of the *Fucus*, from three-fourths of an inch to one and one-half inches in diameter. Examined closely it will be seen to be a dense tuft of stiff, wire-like filaments, many times forked from the base, with wide axils. The apices being nearly all the same length, the tufts look "clipped" all around like a thorn bush. In mounting, it does not adhere to paper. But thinly spread out, in the almost perfect circle which its black frond so naturally assumes, it makes a very pretty appearance on the white paper. It may be found at all seasons and so common that I need not name special localities.

POLYSIPHONIA URCEOLATA, GREV.

The specific name refers to the fruit vessel, which is thought to resemble a little pitcher or jug. The plant is very common throughout the season on the northern shores of both the Atlantic and Pacific Oceans.

It is somewhat variable in appearance, yet when once seen, it is ever afterwards easily recognized. The filaments are much finer and sotter than in the last species, and grow in a loose tuft, four to eight inches high. When taken from the water the plant is flaccid and silky, with a deep, full, rich red color. But when mounted on paper, dry, the filaments are rigid and bristly to the touch, and turn to a dark brown or black with a reddish shade, generally, in places, or over the whole plant. The main stems are from one to three times the thickness of a human hair. They are much branched. But the branches, though somewhat spiney below, do not themselves branch till they have attained a considerable length, when they divide and sub-divide rapidly, making the upper portion of the frond assume a dense and bushy look.

In spreading out on paper, it naturally takes a fan-shaped outline, with a tendency in the main branches to separate from each other, and in the finer varieties to appear twisted. When dried and pressed, there is often a glossy and silk-like appearance to the specimen.

The variety *formosa* is really very beautiful as
its name implies. It is distinguished from the typical
form, by its much finer and silkier filaments, and
by its retaining its rich, red-brown color when dried
on paper.

The open variety, *patens*, is not uncommon, is
more rigid than the typical form, and its end branchlets are recurved. The species grows on rocks,
and sometimes on the stems of *Laminaria flexicaulis*,
in pools, and not far below low tide. I found it
very plentiful in July and August, floating in the
sea, by the rocky shore at Clifton, Marblehead,
and took scores of fine specimens, including every
variety of form. I have some exquisite plants of
the var. *formosa*, taken by my friend, A. R. Young,
at College Point, L. I., as early as May 6th.

POLYSIPHONIA HARVEYI, BAIL.

This is a common and very distinct species. I
have found it in our northern waters, growing most
commonly upon *Zostera*, or "Eel-grass." In the water
it has a marked bushy, or shrub-like aspect, with
stiff branches spreading out widely in every direction,
so that the plant makes a globose outline.

Each tuft is a single frond, stout at the base,
as thick as a bristle, but the parts gradually atten-

uating as they branch. It grows to the height of
from one to three inches, and sometimes more. I
have found it at Wood's Holl, five inches high. It
is invariably dark brown or black on paper, does
not colapse when taken from the water, and is
covered pretty thickly, main stem and branches,
with thorn-like, simple or branched spines, one-tenth
of an inch or less long. The *arietina*, or "ram's
horn" variety, has the end branchlets and spines
recurved or hooked. At Peconic Bay, Harvey says,
the natives call this variety "Nigger hair." I have
found the common form plentiful at Silver Spring,
Providence River, Wood's Holl, and Marblehead, in
July and August. Miss Booth reports it at Peconic
Bay, in September. Mr. Collins, at Lynn beach, on
Zostera, as late as October, and Mrs. Davis finds
it all summer in the "Mill Pond," at Gloucester.

POLYSIPHONIA OLNEYI, HARV.

It is agreed by Dr. Farlow and Prof. Eaton
that this is but an extreme variety of *P. Harveyi*,
and Dr. Farlow is of the opinion that both species
are identical with the older European species, *P.
spinulosa, Grev. P. Olneyi* differs from *P. Harveyi*, in
being a somewhat larger plant, composed of much
softer, and finer filaments, longer and straighter

branches, often with a very decided and sometimes even brilliant pink color, though the more common color is purple brown. It is common in Long Island Sound on *Zostera*, and Dr. Farlow gives the popular name for it there as "Doughballs."

POLYSIPHONIA VARIEGATA,* AG.

This plant has something the same habit as *P. Olneyi*, only that it is larger and more robust, growing often to the height of six to ten inches. Starting at the base with a filament no thicker than a bristle, a half an inch up, it divides into two or more widely spreading branches. These again divide in the same way into long unclothed branchlets. Within an inch of the extremity of the frond, sometimes half way back, all the branches rapidly divide, into long, silky filaments, of a light brown color. The normal appearance of the plant on paper, then, is that of a quarter or third segment of a wheel, with the bare spokes radiating to a rim an inch or so wide, sometimes half the width of the frond, which is made up of these brown pencils of fine capillary filaments. It is quite unmistakable when once seen. It grows parasitical on *Zostera*. It is said to be a winter plant in

* Variegata = Variegated or parti-colored.

19

Charleston Harbor, South Carolina, but is found common along the southern shores of New York and New England in summer. I found it abundant in Providence River and at Onset Bay, and once in Danversport, Mass., the only time, I believe, it has ever been seen growing north of Cape Cod.

POLYSIPHONIA ELONGATA,* GREV.

The three *Polysiphoniæ* to be next described have, according to the books, so many points of resem-blance that you will be at a loss to distinguish them apart if you depend upon the technical account which the books give. And yet, when you have once seen them, side by side, you will never again have any difficulty in recognizing them, and you will wonder why it is that written descriptions cannot make clear differences which are so obvious to the eye. The color of the three is much the same, running from a dark brown, in old specimens of *P. fibrillosa*, through several shades of light brown to a pink in some plants of both *P. violacia* and *P. elon-gata*. I will try to point out the distinguishing marks of the latter species, *P. elongata:*

1. The main stem is robust, cartilagenous, coarse as a pack-thread, and under the pocket lens visibly

Elongata = Elongated.

Jointed in the upper half, as are also all the branches. Sometimes there is a main leading stem and sometimes not. The branches are irregularly placed, but divide and sub-divide in a manner between forking and branching. 2. The axils of the sub-divisions are narrow, so that the branchlets seem to cluster together. 3. Owing to the great length of the secondary branches and branchlets, the plant gives the impression of reaching out and trying to extend itself. 4. The branches seem to maintain their original thickness almost to the tips. 5. On the ultimate branchlets will be found many short ramuli, which taper to base and apex like those of *Chondriopsis tenuissima*. 6. Growing mostly through the same regions as *P. violacea*, it is yet, as compared with that species, if not distinctly rare, certainly very infrequent.

The winter form of this plant, when the finer branchlets are fallen away, is an exaggeration of some of its summer aspects. The great length of its bare, slender, unclothed branches gives it a peculiar and really uninteresting appearance. In this state the natives call it " lobster horns," or " lobster claws," because of its supposed resemblance to the long, slender *antennæ* of that creature. The winter plant very imperfectly adheres to paper.

This is a deep-water species, and is reported as not common all along the coast from New York to Gloucester.

POLYSIPHONIA VIOLACEA, GREV.*

This is by far our most common *Polysiphonia,* considerably outranking even *P. urceolata.* It grows everywhere on the rocks and on several other Algæ, in pools and in deep water, as well as just below tide. I take it often as it comes in upon the waves, with my long-handled dipper, picking out the plants I want, from among the hundreds which go floating by, up and down.

The stem is once or twice as thick as a bristle. Beautiful plants may be found, not more than two or three inches high; but plants from twelve to eighteen inches high, are by no means uncommon.

The distinguishing marks of the species are mainly these: 1. The presence of a leading stem, branched all around in all the fronds. Sometimes there will seem to be two or three main stems. But this appearance arises from the extraordinary development of some of the lower branches. 2. The form of the primary branches, which are long and somewhat widely spreading at the base, but become regularly shorter

* Violacea = Violet colored.

towards the top of the plant. 3. The secondary and remaining branches, which are short, alternately much divided and subdivided again and again, until they terminate in very slender ramuli, which form feathery brown and sometimes violet tufts at the ends, constituting the chief beauty of the plant. 4. Consequent upon this method of branching, the plant has a marked tendency to assume perfect arborescent forms. I have plenty of plants a foot or more high, which almost exactly resemble the great oaks and maples of the forest, and others which are perfect miniature images of the firs and pines, with their regular, tapering, cone-like outline. Our figure in Plate VI., which is a very perfect copy of a plant in my herbarium, could easily be mistaken for a good picture of a forest tree. 5. The stem and main branches are inarticulate. 6. The universal distribution and great plentifulness of the species along our whole eastern coast.

It is an extremely variable plant, and yet the type seems to be as well adhered to as in most Algæ. Many plants, especially those growing in deep water, are very robust and bushy. On the whole it is our most interesting and beautiful Atlantic *Polysiphonia*.

POLYSIPHONIA FIBRILLOSA, GREV.

This is by far the rarest of this group of *Poly-*

siphoniæ. If found at all north of Cape Cod, it must
be very rare. I found some good specimens of it at
Wood's Holl, the last day of July, and Dr. Farlow
reports it at Newport, and Noank, Conn. Miss Booth
at Orient Point, and in Long Island Sound. It is a
summer annual, and grows in deep water, from three
to six and eight inches high. The main stem in the
larger plants is as thick as a pack thread at the base,
but it is soon lost in the multitude of long, large
spreading branches, which it throws out on every side,
so that there is no leading stem in this as in the last
species. The primary branches are long and are themselves irregularly and profusely branched, into secondary
branches, which are much shorter. These again branch
in the same way, and the tertiary branches are usually
covered with spines, not unlike those of *P. Harveyi.*
But the spines are clothed with a dense growth of
colorless fibrills, so fine as to be individually almost
or quite invisible, but in the mass, border all the
branchlets, as they are displayed on paper, with a
light brown "halo" or "mist." This is the characteristic point, and will identify the plant unmistakably,
for it is almost always present. The plant gets its
specific name from these fibrills. The color of the
plant ranges from a light to a dark brown, often even
to near a full black. In general appearance the plant

is not unlike an enlarged, exaggerated, and very spiney *P. Harveyi.* Unlike *P. elongata,* the branches are robust, somewhat bent at various sharp turns and angles, and the parts rapidly diminish in size from base to apex, as they throw out branches and branchlets.

POLYSIPHONIA NIGRESCENS,* GREV.

This is an extremely variable plant, not uncommon along our whole east coast, and identified by one or two distinguishing marks. It is a perennial and grows in rock pools and deep water. It is almost quite black, or very dark brown, when mounted and dry. It has a leading stem, though this is not always easy to make out ; it may, however, usually be detected, as more or less prominent. It is not commonly larger than a bristle. A microscopical dissection of it, shows it to consist of from twelve to eighteen tubes, arranged around a central tube, a singular diversity of habit in a species whose generic congeners are generally so constant to their type, in this respect. Harvey says the best general marks of the species are its many tubed internodes of moderate length, easily visible with a lens ; and its decompound regularly pinnate method of branching. The branches divide and subdivide, alternately twice or thrice in a very regular

* Nigrescens = black.

way. This constitutes the chief beauty, as it is the most conspicuous peculiarity of the plant.

The ultimate ramuli of the young plants, and of the young parts of the old ones, are apt to be *fibrilli-iferous*, in a manner not unlike *P. fibrillosa*, but the method of branching and the general aspect of the plant will easily distinguish it from that.

It is reported all along the coast from Halifax to New York. Miss Booth found it rare at Peconic Bay. I found many specimens of it at Wood's Holl, but took none at Newport, all summer. During several years' collecting at Marblehead, I do not remember to have seen it there, though Mr. Collins finds it abundant along that coast, and Mrs. Davis collects it all summer, on Canal Beach, Gloucester.

POLYSIPHONIA BAILEYI, AG.

The three following California members of this genus which I shall undertake to give an account of, I have put by themselves, not on account of natural affinity, but for convenience of describing them. And yet they are not far apart in the natural system. This is certainly a very distinct and well marked species, like *P. fastigiata*, one which when once seen can never be forgotten, or be henceforth unrecognized.

It grows from three to six inches high, the stem

at first nearly round, more than twice as thick as a
bristle, soon flattened and then immediately and
irregularly much branched. All the branches spring
from the edges of the flattened stem, and the branches
themselves being flattened in the same plane with the
stem, and, giving out branchlets along their edges, the
whole plant is built up in one plane. The main
branches spread widely, and are irregularly placed.
But the secondary branches are very regularly alternate,
the one-tenth of an inch or so apart. Toward the
base of the branches, in all the old or full grown
plants, these branchlets will be found broken off,
leaving nothing but short stumps. The branchlets
themselves consist of a short stem, one-eighth to one-
half an inch long, clothed on each side and at the
top all around with very short, alternate simple or
compound awl-shaped, incurved ramuli. These branchlets are generally about the same length along the
sides of the branches, but here and there one will
shoot out beyond the others, and sometimes it will
put out branchlets like a primary branch.

Dr. Anderson reports it scarce at Santa Cruz, on
rocky beaches, all the year around. Mrs. Bingham,
and Dr. Dimmick, find it very common, thrown up
on the beach, and growing on small rocks, in all
seasons, at Santa Barbara. Mr. Cleveland reports it

common at San Diego. It is among the most common forms that come to me from my correspondents
on the Pacific coast. The color is a full black. It
adheres very imperfectly to paper.

The artist has very excellently represented a frond
of this species, in Plate VIII.

POLYSIPHONIA PARASITICA, GREV.

This species in many respects, and especially in
general aspect and outline, resembles the last, but
differs from it by being smaller, of a much finer
and more delicate substance, and lighter color, which
is usually a light reddish brown. I have never seen
typical forms of this species over two inches high.
The figure in Plate VII, excellently well pictures
not only the color but every characteristic feature of
this very beautiful plant. The stem, branches and
branchlets are all flattened and branch from the two
edges, primary branches irregularly and very widely,
secondary regularly, widely, alternately. The secondary
branches are mostly little plumes, or themselves
bearers along their edges of little plumes. The
branching of all the small parts, even to the
minutest, is regularly alternate. This gives the plant
a very delicate, feathery appearance, very greatly like
the finer fronds of *Ptilota plumosa.* My correspond-

ents report it extremely common in Southern California, but somewhat rare in the north, growing upon
the large rocks and upon other Algæ, and in tide
pools, all the year around.

Variety *dendroidea*, differs more in appearance
from the normal form than do some fully differ-
entiated species, and yet, after a careful examination,
you will find that the difference consists fundamentally
in the branching being made at a much more acute
angle in the variety than in the typical form. The
frond stretches out to a considerably greater length,
four or five inches sometimes, "long, slim and slender"
in appearance. The main branches are placed at irregular intervals, but the secondary, at regular intervals,
alternate. From the extreme narrow angle, at which
the parts branch, they all appear to hug close to the
main stems, which gives the slender, narrow look to
the frond, and effectually prevents the beautiful
plumose aspect, which is seen in the whole plant,
and in its smallest parts, in the normal form. The
color of this variety is a full black, or a very dark
brown. In the young parts of both varieties, the
interior joints of the fronds may be easily seen with
a pocket lens. This variety seems to be even more
common along the whole coast than the normal form.
It does not adhere to paper.

Polysiphonia Woodii, Harv.

Although this plant seems to be built on the same
general plan as the other two California species,
already described, it is yet sufficiently distinct to be
not only a good species, but also easily recognized.
The stem is, perhaps, twice the size of a bristle,
divided from near the bottom into long, spreading
branches, the whole plant being from four to six
inches high. All the parts are flattened, the younger
visibly articulate, and branch from the edges in one
plane. The secondary branches also separate with
wide axils, but give out their branches at narrower
angles, while the ultimate, awl-shaped ramuli are much
inclined to be incurved, rarely to spread widely. The
plant varies much in particular respects, depending
much, I find, upon whether it bear the sexual or
asexual fruit, or be sterile ; but the difference usually
consists in the lengthening or shortening of the parts
of the plants, some being thick, dense and bushy,
others slender, spreading and feathery. It is very
common at all seasons, Dr. Anderson says, at Santa
Cruz, growing chiefly on *Macrocystis*, and, therefore,
of course, in deep water. Dr. Dimmick collects it
on the beach at Santa Barbara, and Mrs. Bingham
gets it there, early in the season, upon *Halidrys*,
also. It adheres well to paper and makes, in most

cases, a very pretty specimen. The color is a light brown.

Genus.— *RHODOMELA,* * *Ag.*

RHODOMELA SUBFUSCA, AG.

The *dark brown Rhodomela* is a common plant along our shores, from New York northward. It seems to be quite at home in all northern seas, as it has been found in Nova Zembla, and the Ochotsch Sea, as well as in all northern Europe and America. The ripe, robust, black, typical form is far from handsome ; but the young plants, which go under the variety names of *Rochii* and *gracilis*, are extremely beautiful. It is a perennial, and its winter and summer aspects differ greatly. In the winter all the finer portions of the frond fall away, leaving the long, lateral branches, and the main stem standing stiff, naked, dark and ungainly. But in the spring and early summer, when it is clothed in a new growth of delicate brown branchlets, it is a very graceful and charming plant.

It is found attached, by a thin discoid holdfast, to rocks, stones, and shells, near or below

* Rhodomela Red-black.

low-water mark   The fronds are from six to twelve inches high, cylindrical, as thick as a pack-thread, in full-grown plants, much slenderer in others, fine as thread or hair in young plants, and in var. *Rochii.*   In the common form, the main stem and branches are cartilaginous, stiff, and when dry, hard and harsh, and quite black. From the leading stem, which runs to the top of the plant, the branches spread out on all sides, the lower being the longest, often as long as the main stem — gradually short-ening towards the top. The branches are all more or less naked below.   But, towards the end, they divide and sub-divide rapidly in alternate ramifica-tions, so that the small branchlets are much crowded, and, on paper, the primary and secondary branches seem thereby to terminate in little brooms.

This is true only of the full-grown, typical forms, and of the var. *gracilis*, a most excellent representa-tion of which appears in Plate IX. The normal form differs from this only in being more robust, of a less regular habit, and of a much darker color. The var. *Rochii* is much finer and softer, and the end branches are quite separate, but tipped with a very fine pencil of hairs. This is the early spring form, and is found chiefly south of Cape Cod. I have an exquisite specimen collected by Mr. Young,

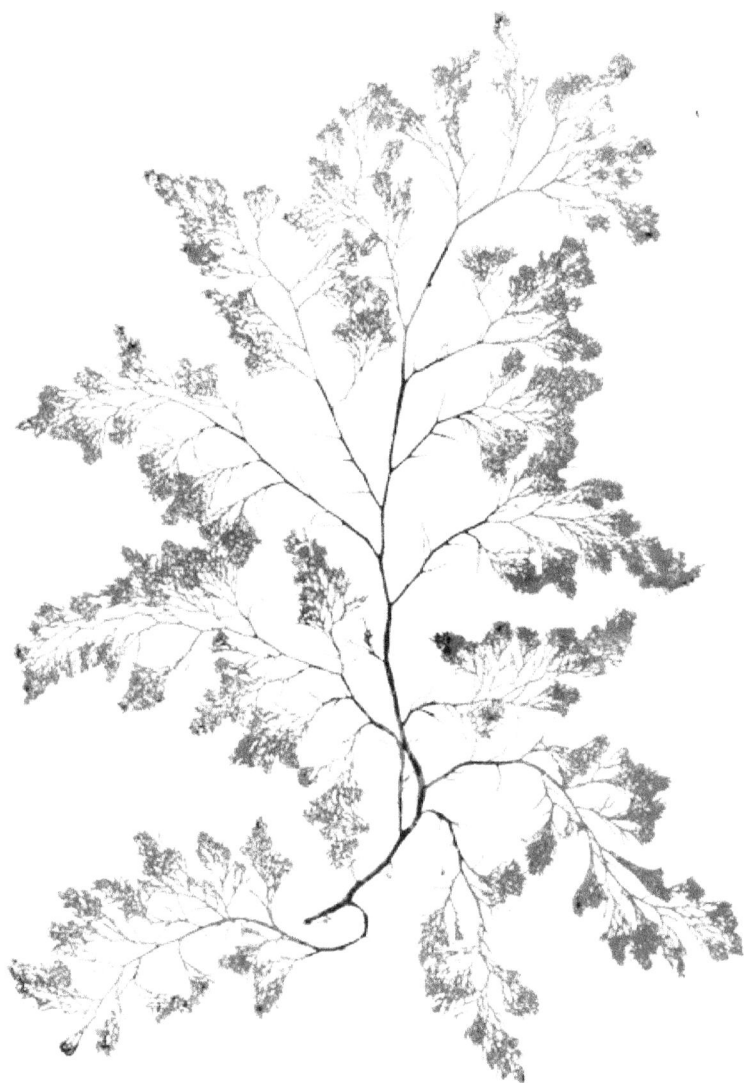

RHODOMELA SUBFUSCA, *Ag.* *var.* GRACILIS.

PLATE X.

of Brooklyn, as early as March 27th. Var. *gracilis*
is more common in our northern waters, and approaches more nearly the typical form. The specimens in my herbarium are of a rich, slightly reddish
brown color. Whoever will take the trouble to look
for this plant in the early spring, will find it one
of the most beautiful of our marine flora.

RHODOMELA LARIX,* AG.

This and the next species grow on the California,
and north western coast. *R. larix* is an arctic
species which has made its way as far south as
Santa Cruz and Monterey, but appears south of there,
only as a rarity. It has been found at Santa Barbara, by Mrs. Bingham, in May; and in January
and March, by Mr. Cleveland, thrown up from deep
water at La Jolla Point, San Diego. It was brought
from Nootka Sound, by Menzies, more than three-
quarters of a century ago, and described and figured by Turner, in his unequalled "Historia
Fucorum." Dr. Anderson reports it as very plentiful at Santa Cruz, and northward, growing there
at all seasons, on the shelving rocks of soft sand-
stone or shale.

The frond is robust, cylindrical, **thick as a**

Larix = Larc

crow's quill, from six to fourteen inches long; at first unbranched, but soon much branched all around, with limbs of various length, which stand out straight from the main stem. Branches from one to four and five inches long, according to the size of the plant.

The distinguishing mark of the species is the presence upon both stem and branches, of little tufts, or clusters of incurved ramuli. They are spirally placed, but when the plant is mounted, they seem to be alternate. They are commonly so far separated as to be quite distinct, and are not more than a quarter of an inch long. Color of the plant when dry, a jet black.

RHODOMELA FLOCCOSA,* AG.

This species differs from the other in many marked points. It is less · robust in habit; the stem and branches are flattened; the whole frond is divided and sub-divided in one plane; the branches are alternately set upon the stem, and once or twice alternately divided; the ultimate ramuli are some-what incurved, but not clustered as in the other species. In fertile plants, the last divisions at the end of the branches are more or less gathered into

---

* Floccosa = Full of locks of wool.

a mass, as in the whole genus, but in a far different way from the thick tufts of *R. larix.* In truth, the plant very much resembles the fronds of *Polysiphonia Baileyi,* for which it will be more often mistaken than for any other species. You will get a good idea of the general appearance of the plant, by consulting Plate VIII. It differs from *P. Baileyi* chiefly, in being somewhat more coarse and robust.

The main stem, in plants four inches high, is not much larger than a bristle. It is found from four to ten inches long. Color, a full black. It grows at Santa Cruz, on the rocks, in the same situation as its companion species, but is much less common, and is collected from September to November. At Santa Barbara, Dr. Dimmick found it common near the lighthouse, and Mrs. Bingham says it is very common there all the year around, growing with *Polysiphonia parasitica.* My specimens from there are mingled with plants of that species.

Genus — *CHONDRIOPSIS,** Ag.*

This genus is represented by three common species on our New England coast, and by one on the coast of California. The Atlantic species all belong

* Chondriopsis = Somewhat cartilaginous.

to the warmer regions, and grow south of Cape
Cod, but grow there in great abundance. Though
not a very striking or beautiful genus, it is yet far
from being uninteresting. It is characterized by two
marks which make it extremely easy of recognition,
viz.: The uniform light or dull brown color when
fresh; and the fact that the stems and branches
are pretty thickly covered with short club or spindle-
shaped ramuli. These ramuli, which are from one-
eighth to one-half of an inch long, are very much
constricted at the base, often seeming to be attached
by the finest thread, or hair, to the plant. In three
of the species they taper to a fine point at the
extremity, and in the other, *C. dasyphylla*, they are
very blunt at the end, shaped not unlike a boy's
top. The plants should not be put in fresh water,
and should be dried under comparatively light
pressure.

CHONDRIOPSIS TENUISSIMA, AG.

This, as its name implies, is the slenderest of
the several species. It grows from four to six
inches high, with an undivided stem once or twice
as thick as a bristle, with long, spreading, mostly
alternate branches, sometimes simple, sometimes themselves branched in the same way, and furnished
throughout, more or less, abundantly with the charac-

teristic ramuli, one-fourth to one-half an inch long, slender and attenuated to a sharp point, both at the top and at the place of insertion on the branch. In drying, the plant adheres well to paper. It grows between tides, on *Fucus* and on rocks. It is a summer annual, inhabiting Long Island Sound and adjacent waters. I have collected it only at Wood's Holl. Miss Booth reports it in great abundance in Peconic Bay.

CHONDRIOPSIS STRIOLATA, AG.

Frond from four to six inches high, twice as thick as a bristle, with a short stem, soon dividing into many long, simple, or once or twice compound branches. The branches rise somewhat perpendicularly, and make a compact tuft of the plant. The ramuli are very plentiful, much constricted at the base, somewhat rounded at the apex; standing near the next species, in this respect, as it does near the last in its slender habit. The ramuli not unfrequently bear like secondary ramuli along their sides. This is the characteristic point in the plant, though it sometimes occurs in *C. dasyphylla.* This species grows on rocks and other Algæ, in pools, between tides, and below. I have taken it, at low-tide, in great abundance, on the rocks, east of the first beach, at Newport, in July and August. It is plentiful at Peconic Bay, and all through Long Island Sound and southward.

CHONDRIOPSIS DASYPHYLLA,* AG.

This is a considerably more robust plant than either of the others already described, growing from six to twelve inches high in bushy tufts, the main stem and branches being as thick as wrapping twine. There seems to be, at least, two distinct types, or varieties, of this species. The one has a pronounced leading stem, with relatively shorter and more erect branches, and the ramuli longer and less blunt, or only rounded at the apex, like those of *C. striolata*. The other just as manifestly divides up near the base into several long, widely spreading, similar branches, which are clothed throughout with an abundance of short, secondary branches. The ramuli of this variety present the typical form, much attenuated at the base, short, thick, very blunt, top-shaped, or truncated at the apex. The former I found very plentiful at Newport, in July and August, growing in rock pools, near low-tide, and, as it lies pressed on paper before me, presents a mixture of green and purple color. The latter was among the most abundant of the plants in the little harbor

---

* Dasyphylla = With bushy foliage.

at Wood's Holl, the last days of October. In the water it was olive, but in drying it turned black.

Chondriopsis nidifica, Harv. ♦

This plant is a native of the Pacific coast. It grows to the height of six or eight inches, as thick as a sparrow's quill, cylindrical, inarticulate, sparingly branched, in a manner between alternate and forking. Branches several inches long, quite simple, or once or twice forked. The branches are either altogether naked, or bear, at considerable intervals, little tufts of short, incurved fruit-bearing ramuli, a quarter of an inch or so, long. This is the distinguishing feature of the plant. I have plants, but no notes of this species, from my correspondents on the Pacific coast. Another species, which Agardh reckons the same as this, *C. atropurpurea*, is also found on that coast. I have specimens, but no data for telling how plentiful it is, or where it may be found.

Genus.— *LAURENCIA.** *Lam.*

But three species of this genus are reported on the California coast, two only of which are sufficiently common to come within the scope of this book.

---

* Laurencia.— Named for M. de la Laurencie.

LAURENCIA PINNATIFIDA,* LAM.

Frond, flattened, narrow, in specimens ten inches long, not less than one-fourth of an inch wide; substance cartilaginous, thick; color a livid purple, becoming brownish in drying, and often faded to every shade, down to a dull white, and not seldom so unevenly faded, that you will get every sort of color in the different parts of the same plant. The frond widens somewhat upwards, and the flattened branches are often as wide as the main stem. The stem is usually naked at the base, owing, no doubt, as the appearance indicates, to the breaking off of the lower branches. An inch or two above the base the branches appear upon the edges of the flattened stem, opposite or alternate, at an angle half way from horizontal to perpendicular. The branches themselves are branched in the same way with flattened branchlets along their edges, and in rare cases these again. The plant is never more than three times pinnatifid, rarely more than twice. The ends of the ultimate pinnulæ are always quite blunt.

The points indicated above will easily identify it. Dr. Anderson finds it growing on *Laminaria*, not uncommon, at all seasons, at Santa Cruz. At Santa

* Pinnatifida = Pinnately cleft.

Barbara, Dr. Dimmick and Mrs. Bingham find it growing near low-tide, and in deep water, upon the rocks, from which it is thrown upon the beach. Mr. Cleveland gives substantially the same account of its habit at San Diego, where he collects it from November to March.

LAURENCIA VIRGATA,* AG.

This species has much the same geographical range as the last, but is not so common, I judge, from the comparative infrequency with which specimens find their way to the Atlantic states. It differs also, in being cylindrical in stem and branches, and by having the branches set all around the stem, and not on two sides only. The general habit of the branching, except as to that, is much like the last. In size, substance and color it greatly resembles *L. pinnatifida.*

Order.— CHYLOCLADIEÆ.
Genus.— CHYLOCLADIA,† Grev.

The only plant which later revisions have left in this genus from our flora is the one which both Harvey and Agardh call *Lomentaria ovalis.* But as it has been lately known, and distributed, among American

---

* Virgata, refers to its long, rod-like, branches.
† Chylocladia = Juicy-branched.

botanists, under the generic name given above, we will adhere to that.

CHYLOCLADIA OVALIS, HOOK.

The frond is cylindrical, as thick as a goose quill, six or more inches high, forking and sparingly branched ; the stem and branches are densely clothed near the summit, with ramuli, which resemble little sacks or bags, from one-fourth to one-half an inch long, sometimes shaped like an Indian club, and sometimes like an egg, hence the specific name. It grows, Mr. Cleveland says, in deep water, and is collected as a rare plant at Point Loma, La Jolla, between December and April. The var. *Coulteri*, is among the most common of plants at Santa Barbara growing on rocks at mid-tide and in deep water. It is not rare at Santa Cruz, where Dr. Anderson finds it on the sides of soft rock cliffs, near low-tide. It is not found on our Atlantic shores.

Order.— *SPHÆROCOCCOIDEÆ.*
Genus — *GRINNELLIA,* Harv.*

GRINNELLIA AMERICANA, HARV.

Somebody says, " Doubtless God could make a better fruit than the strawberry, but doubtless He

---

* Grinnellia.— Named for Mr. Henry Grinnell, New York city.

never did." So may we say of this Alga, "Doubtless
the Hand that fashioned this graceful and brilliant
plant could make a finer. But it is certain He never
has, to grow on our shores, at least."

Holding to stones and shells by a minute disk, not
so big as a pin-head, with the merest thread of a
stem, not a quarter of an inch long, it grows down
on the sea bottom, five or six fathoms deep. From
this slender thread of a stem, the wavy-edged, thin,
delicate red membrane of a frond, gradually expands
to the width of three or four inches, and rises to the
height of one to two feet or more, tapering to a
rounded point at the top. Along the middle of the
whole length of the frond, runs a fine but distinct line
of deeper color, and apparently thicker substance,
which not a little resembles the midrib in the leaf of
terrestrial plants. The edges are full, and ruffled,
or wavy, so that when put on paper they fold in
"plaits," at regular intervals, deepening the color at
these places, and adding another charm to the picture
which the mounted plant makes.

This beautiful plant grows along our shores from
Long Island Sound to Fortress Monroe, being most
abundant and most luxuriant about New York Bay.
It is in its perfection by the first of August, when it
loosens in great numbers, from its deeper fastnesses,

and floats to the surface, and is driven in shore. Then like Macbeth's bloody hand, it almost seems the

> "Multitudinous seas to incarnadine,
> Making the green — one red."

There lies before me as I write, half a dozen splendid fronds taken at that season, on the pebbly beach, where the Hessians landed at the battle of Long Island, just below Fort Hamilton, New York. They are from one and one-half to two and one-half feet long, and three to four inches wide, perfect in outline, and of a most beautiful rosy red, with just a shade of orange here and there. They would make exquisite pictures framed as pannels. A reduced copy of one of them adorns this volume, in Plate XII. They are delicate plants, and must be treated tenderly, and yet these specimens were carried, rolled up in newspapers, from New York to eastern Massachusetts, 250 miles, and kept twenty-four hours out of water, before they were mounted.

Genus.— *DELESSERIA,* Lam.

DELESSERIA SINUOSA, LAM.

The *Delesseria* with a sinuous or indented outline is a deep water plant, growing on the roots of *Lam-*

* Named for Delessert, a French botanist.

inaria flexicaulis, and on shells and stones, at a depth
of ten to forty fathoms. It has been collected on
the coast of Maine at a depth of seventy-five, and
in the Arctic seas at a depth of eighty-five fathoms.
It is very plentiful in Massachusetts Bay, and along
the whole coast northward. It is sparingly found
soutn of Cape Cod. It is to be looked for among
the masses of sea weeds rolled up by the tides along
our northern — especially rocky and pebbly — beaches.
It is scarcely ever absent from such *rejectamenta*
of the sea, for it is a perennial. It is as easily distinguished there, as are the leaves of the oak or
maple, among the fallen foliage of the forest. In
some of its forms, it bears no inapt resemblance to
the young leaf of the oak. In England, it is called
the oak-leaf *Delesseria.* In California, we have the
true oak-leaf form, called *D. quercifolia,* which is
not much unlike this species.

The plant grows from three to six inches or more
high. It is sometimes narrow, and sometimes quite
broad as is the one, which is copied for this volume, and represented life size, in Plate X. It is extremely variable in outline, but the fact that it is
the only red Alga which has a regularly midribbed and
veined frond, like the leaves of trees, removes all
difficulty in the way of its ready recognition, when-

ever it is seen. Its color is a deep lake-red, when fresh or young, but often flecked with green, or white, or yellow, or faded to pink, when it has been long exposed on the shore. There are very many beautiful plants to be found among its various forms. It does not readily, or very firmly adhere to paper in drying.

I find, from an old work on my shelves, by Gmelin, of St Petersburg, that it was described more than a century ago, he having then already, received specimens of it from Kamtschatka. It is essentially an arctic plant. I have two copies from Spitzbergen, where it is described, as being among the most common of the red Algæ.

DELESSERIA ALATA, LAM.

The *winged Delesseria* has the same general habit as *D. sinuosa*, except that it is a very much narrower, and more delicate plant. It grows in much the same situation, and may be looked for in the same places. It will almost always be found on our shores in connection with *Ptilota plumosa*, var. *serrata*, on whose frond it is very commonly parasitical. It is commonly not more than three inches high, though I have both English and American specimens, twice that.

The cylindrical stem flattens into a midrib, directly it enters the leafy part of the frond. There is but a very narrow margin of leaf, or wings, bordering the midrib; in our plants, it is not over one-eighth to one-fourth of an inch wide. The frond rapidly forks or irregularly divides, in one plane, so that the frond has a multitude of narrow, terminal ramifications, along towards the end of which, the midrib, in most of our American plants, seems to disappear.

The margins of the lobes are usually entire, and they run out commonly to a narrow, but *always rounded*, termination, nearly one-tenth of an inch wide.

It will often be found associated with *Euthora cristata*, from which it will sometimes be found difficult to distinguish it, on account of similarity of size and ramification. But the small ends of the *Euthora* are *never rounded*, but always *square* or *notched*, in an angular fashion. A common pocket lens will always reveal the distinction, if it cannot be made out with the unaided eye. *D. alata* is a perennial. It has not been found south of Cape Cod, but it will seldom be wanting on our northern shores. It is not uncommon on the California coast. Its color is a light red or delicate pink. It is indeed a very beautiful plant when carefully mounted. Our American plants seem to adhere well to paper.

Genus —*NITOPHYLLUM*,* *Grev.*

This splendid genus must be one of the glories of the marine flora of California, a coast extremely rich in fine and beautiful species. With its many species of large and brilliantly colored plants, thin and silky in texture, graceful in outline, prolific in numbers, surely this genus would be difficult to match. What could be more charming than a wide, deep, clear rock-pool, where the brown " Kelp " and the green *Ulva*, intermingled with the waving fronds of these crimson plants, should spread themselves out in calm and lazy life, the wonder and admiration of every beholder ; or to look

> " Far down in the green and glassy brine,
> Where the floor is of sand, like the mountain drift,
> And the pearl shells spangle the flinty snow;
> Where from Coral rocks the sea plants lift
> Their boughs where the tides and billows flow."

And see there, growing upon the stems of the giants of Neptune's forest, these brilliant fronded *Nitopyhlla*,

> · Red like a banner bathed in slaughter?"

NITOPHYLLUM SPECTABILE, EATON.

This truly *admirable* plant, says Prof. Eaton, is " among the largest species of the genus, often two

* Nitophyllum = A shining leaf.

feet long, and in the spread of the lobes two-thirds as broad. The frond has usually a central body with forked, tongue-like, marginal branches, an inch wide and six or eight inches long. The lobes are often crowded so as to overlap each other. No veins are visible. Fruit dots are scattered over the surface of the frond. The substance is rather firm, but thin, and does not very well adhere to paper, except in the younger portions. The color is dull purplish-red, more rosy in the newer parts."

I have seen only small specimens of this noble plant. Dr. Anderson reports it quite common at Santa Cruz, and when he also reports, that three other of the largest species of this splendid genus are among the commonest plants in those waters, I cannot help wishing that that El Dorado of the Algologist were not so far away. He says all the species of *Nitophyllum* grow between tides, on rocks, and on the roots and stems of *Laminaria*, of course in tide pools, all the year round. No doubt they grow in deep water there also, as they do, according to Dr. Dimmick, at Santa Barbara.

NITOPHYLLUM LATISSIMUM,* AG.

The frond springs from a narrow base, and spreads

* Latissimum = Widest.

out widely in lobes, like a hand with the fingers
extended, or remains entire, a foot long, rounded at
top, four or five inches wide, or displays one long,
tapering lobe and several smaller ones by the side
of it. It will thus be seen to be extremely variable
in form. But it has one mark which will infallibly
distinguish it, viz.: a network of branching, crossing
and interlacing veins, which covers over the entire
frond. The veins are very pronounced, and about
equally so throughout the frond. At least one other
species, of this genus, from these waters, has veins in
the frond, viz.: *N. Ruprechteanum.* But they
are mostly parallel, and rapidly fade out as they get
to the middle of the frond. Mrs. Bingham and Dr.
Dimmick find it not very common at Santa Barbara,
thrown up from deep water, in May and June. It
does not occur at San Diego. Dr. Anderson's report
of this and other *Nitophylla,* is given under the last
species, *N. spectabile.*

NITOPHYLLUM, FRYEANUM, HARV.

This plant was no doubt named for Mr. A. D.
Frye, of New York city, one of the earliest collectors
of Algæ on the Pacific coast. It is neither a large
or a very common species. It attains a height of
five or six inches, and is spread to about the same

width when full grown, and much divided. From
a minute point of attachment it widens rapidly upward
in a wedged-shaped manner, quite like a palmate, or
typical form, of "Dulse," and in general, it may be
said to have the habit of the smaller species of that
genus, found in the same neighborhood, viz.: *Rhodymenia corallina*. The full grown frond is divided
almost to the base into three or four lobes, and these
again at top, having widened much, are themselves
divided half way down, the secondary lobes being
nicely rounded and scalloped at top. It is full red,
thickish and nerveless. It is not very uncommon in
northern California, but is rare in Santa Barbara, and
has not yet been found at San Diego. In the former
place it is thrown up from deep water in May, and
probably at other times.

NITOPHYLLUM ANDERSONII, AG.

Though by no means the largest, this is one of the
most interesting and certainly the best marked species
of the group. It has a narrow frond throughout, not
over one-third of an inch wide, often less than that.
It throws out branches profusely along each edge,
or quite loses itself in branchings and forkings, so as
to make often, a very rambling and uncertain outline.
But the figure, in Plate XI., will give a much better

12

idea of the plant than can be conveyed by any words. It has one unmistakable mark which will distinguish it from every other member of the family, viz. : the fact that all the parts and lobes are armed along their edges with sharp, forward-pointing teeth. In all the older parts, a midrib is very distinctly seen, which loses itself at last near the middle, or toward the younger parts of the frond. My largest specimens are eight inches in lateral spread, and something less in height ; color, a dull or brownish red. It is common along the whole coast, and at Santa Barbara, it is reported growing in deep water near the wharf, and on large rocks at low-tide, and at San Diego, in deep water, from November to April.

NITOPHYLLUM RUPRECHTEANUM, AG.

This is a fine, large and well marked species. Starting from a narrow stem, it soon expands into a repeatedly forking, widely spreading frond from one to two feet· long. The strap-like lobes of the frond are from half an inch to one inch wide, of various lengths, of nearly parallel edges, rounded and often cleft at the top. The edges of all the older parts of the frond, and of any old breaks in it, are bordered with a fringe of minute leaflets, not more than one-eighth of an inch long. Sometimes these extend over

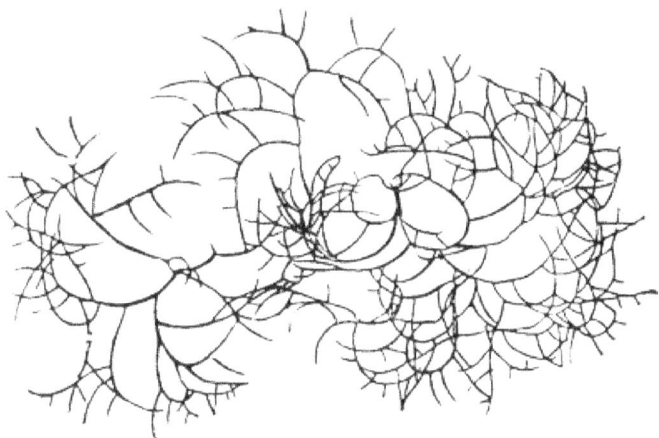

LOMENTARIA BAILEYANA, *Harv.*

portions of the surface of the frond. This is an unmistakable mark of the species. The thickened stem divides and forms midribs or veins in the lower divisions of the frond. These, however, soon disappear upward. The color is a dark red with a shade of purple. Substance, somewhat rigid. It does not adhere well to paper. It is among the commonest of plants along the whole coast, and must be one of the finest features of a fine flora.

NITOPHYLLUM FLABELLIGERUM,* AG.

This is another large plant growing a foot or more high, and spreading as wide. In general habit it very much resembles the last species, but differs in lacking the fringe of minute leaflets upon the edge of the lobes. It is also more widely divided in the palmate frond, the lobes are more numerous, more wedge-shaped, shorter and narrower. From a flattened stem, one to four inches long, the frond spreads, by repeated forkings and dividings, into many segments with rounded tops. Large, dark, fruit dots are scattered over the surface of the fertile fronds. It appears to to be a native of the northern shores, as I have not received it from any locality south of Santa Cruz.

* Flabelligerum = Fan-shaped.

NITOPHYLLUM VIOLACEUM, AG.

This species is distinguished by its very narrow frond, which forks almost from the bottom, into long, slender segments, and by its marked purple or violet color. It is quite a variable plant, yet one or the other of these marks will usually determine it. It grows to the height of six or eight inches, and its lobes are often not over a quarter of an inch wide, and are apt to throw out at irregular intervals along the margin, minute leaflets with a dark spot in them ; this is the fruit. It is plentiful along the entire coast, and grows in deep water on the larger Algæ.

Genus.— *CALLIBLEPHARIS,** *Kutz.*

CALLIBLEPHARIS CILIATA, KUTZ.

The *ciliated* species of this genus is by no means as common in our waters, as it is reported to be on the other side of the Atlantic, but it will well repay looking for where it may be expected. It is an annual, growing in deep water, and ripening its fruit and frond in early winter. It is found at Cape Ann, and down the coast of New Eng-

* Calliblepharis = Beautiful eyelashes.

land and the Provinces, as far as Halifax. Mrs.
Davis gets it on the beach at Gloucester, where it
is thrown up, from September to December. Prof.
Eaton found it at Eastport, Me. It may be expected at all intermediate points.

It grows from a mass of short, creeping roots,
at first, a short, cylindrical stem, which gradually
expands into a flat, thickish, cartilaginous frond,
from one-half to one inch wide, and from two to
six inches high, tapers again at the top into a simple acute apex, or, forking, ends in two such apices.
Along the edges of this frond, at irregular intervals,
there come forth, at first, sharp, minute, spine-like
processes, usually curved. These at length grow
into miniature fronds of the same general form as
the parent frond. These again put out the *spinous
cilia* ("eyelashes," so called) which, in turn, become still more minute fronds, of the original pat-
tern, having ciliated edges. Here, generally, the
ramification stops. The plant has a clear, strongly
marked red color, with a decided tendency to turn
darker in drying. It adheres well to paper.

Genus.— *GRACILARIA,* Grev.

GRACILARIA MULTIPARTITA, AG.

The *many-times-divided Gracilaria* is the only representative of this genus, which grows in our northern waters, and it is found on both the east and west coast, being quite common in Southern California. The narrow form, *angustissima*, is very plentiful in Long Island Sound and adjacent waters. I have collected this variety in considerable quantities in Providence river, in the month of August, where Prof. Bailey and Mr. Olney found it in abundance, many years ago. It has been reported north of Cape Cod, by but one collector, Mr. Collins, who finds it quite plentiful in the warm waters and on the muddy bottoms of Mystic river marshes, near Boston, from May to November.

The plant is an extremely variable one. It grows to a height of from six to twelve inches. It starts with a short, cylindrical stem. This immediately begins to flatten, and directly expands into a narrowish flat frond, which *always widens upward*, till it is a third or half an inch broad. Then it divides into two to four segments, which are, in the same way,

* Gracilaria = Slender, graceful.

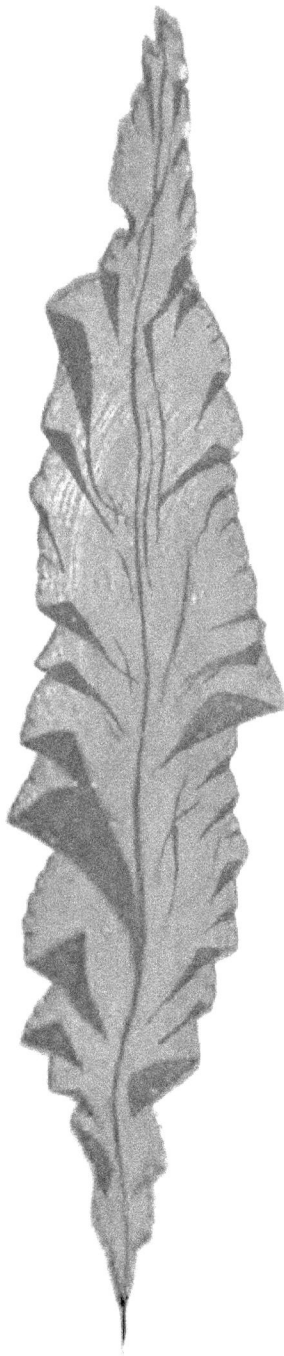

GRINNELLIA AMERICANA, *Harv.*

slender at first, but gradually widen as they grow upward. Another division, soon occurs in each of these, and the parts again expand, and so on. This method of growth, together with the partings or branchings which occur along the edges of the frond, and which likewise have the same habit of upward widening, gives the whole frond a decidedly fan-shaped aspect.

In July or August, the seed-vessels appear along the edges of the branches, like warts, as big as pigeon shot. The substance of the frond is somewhat tender and brittle, but when dry, it is tough and leathery. The color is a dull purplish-red, but much darker when dry and mounted on paper, to which it adheres rather imperfectly.

Order.— *CORALLINEÆ.*
Genus.— *CORALLINA,* * Lam.*

There are several genera of this order growing on our shores, besides the one named above. They are all characterized by the calcarious, or stony incrusta tion of the frond. Some of them are mere pink or brown patches, upon the fronds of other Algæ, or

* Corallina = A litt'e coral.

upon the rocks, stones and shells; others grow up
in the form of plants. None of these, with the exception, possibly, of the *Corallina*, and the *Amphiroa*, will be of sufficient interest to any other than
the scientific botanist, to make them desirable to
collect. But that you may know, that these things
which you will find so plentiful all along the shore,
and which much more resemble, by reason of their
stony structure, the corals than any plant, are real
plants and not corals, I have selected one species
for description. It should be added, perhaps, that
the true plant structure, and the reproductive organs,
really exist as in other red Algæ, but are concealed
beneath the hard crust which is secreted upon the
outside.

CORALLINA OFFICINALIS, L.

The *medicinal* species of this genus is the only one
on our eastern shore. It is also a native of California. It grows in great abundance in tide pools,
and upon the rocks, about low-water mark, all along
our shores from New York northward. It is from
one and a half to three inches high, extremely variable in size and aspect, in some cases loosely and in
others densely tufted; in color, from a reddish purple to a gray green, and if exposed to the weather,
for a little time, upon the beach, bleach out quite

white. The frond is composed of cylindrical filaments, a trifle flattened, the main stem branching from its edges, as do also the principal branches. The whole plant is built up of small stony, somewhat wedge-shaped joints, a trifle the widest at the top, all the branches and branchlets spring from the top of the joints directly below. It generally refuses to adhere, but may be fastened down with straps of gummed paper.

Order — *GELIDIEÆ*.

Genus.— *GELIDIUM,* * *Lam.*

One species of this is a native of both shores, and the others of the Pacific alone. They are narrow, compressed, rarely quite cylindrical plants, of a firm, tenacious substance, and, when dry, quite rigid and horny. They are pinnately branched, and the branching is mostly in one plane.

GELIDIUM CORNEUM, LAM.

This is a most variable plant. A typical form, such as we figure, in Plate XIII, will not very frequently be found. But every plant will be but a variation on that theme. Plants of this species on

---

* Gelidium = Ice-like or jelly-like.

the eastern coast are small, not more than an inch, or an inch and a half high. Those growing in California are three or four inches high, the lower branches long and naked below, gradually shortening toward the top of the plant. They are two or three times pinnated, that is, the branches bear branches, and these branchlets, arranged on the same pinnate plan throughout ; the ultimate ramuli are usually club-shaped, and swollen with the spore masses, which they contain. Color, a purplish red, but by exposure on the beach, it fades through all shades to dirty white. It grows in tide pools on rocks and other Algæ, near low-water mark. It is extremely common on the Pacific coast at all seasons. A section of the fruit-bearing branchlet makes a very interesting microscopical object, with its club-shaped spores, growing from a central partition, which divides the inner cavity of the conceptacle into two equal chambers.

GELIDIUM CARTILAGINEUM, GREV.

The fronds often attain a height of twelve inches, are flattened, two-edged, one-tenth of an inch in diameter, flatter upwards, three or four times pinnated. The root is a mass of much-branched, rigid fibres. Stem and long primary branches naked be-

GELIDIUM CORNEUM, *Lam.*

PLATE XI

low, thickly, pinnately branched above. All the
lesser pinnules issue at very obtuse angles with
distinctly rounded axils. Color when growing is a
very dark purplish-red. Its size, the long primary
branches, and the rounded axils of its ultimate
branchlets, distinguish it from the last. It is very
common at all seasons, growing between tides, on
rocks and weeds. Mrs. Bingham finds it on the
stems of *Phyllospora Menziesii* at Santa Barbara.
At San Diego it grows in deep water and in deep
tide pools. It does not adhere to paper in drying.

GELIDIUM COULTERI, HARV.

This is much the smallest and most delicate
species of the three. It grows in considerable tufts
from a mass of matted root-fibres, sometimes fifty
plants together. It is very slender and narrow, not
more than the twentieth of an inch wide, yet all
parts are clearly flattened, and the opposite pinnate
branching, goes on very regularly from the edges.
The fronds are commonly two or three inches high;
the primary branches one to two inches long; the
secondary are usually the club-shaped ramuli which
contain the fruit, and are closely set and opposite.
Color, a very dark purple. It adheres to paper fairly
well. Beginning as a somewhat rare plant in San

Diego, it becomes more and more common toward the north. At Santa Cruz it is very plentiful. Its habitat is upon rocks and other Algæ between tides.

Order. — *HYPNEÆ.*

Genus.— *HYPNEA,** *Lam.*

Hypnea musciformis, Lam.

The *moss-like Hypnea* is in many places south of Cape Cod, a very common plant. I collected it at Wood's Holl, but not very plentifully. Miss Booth speaks of it as growing "by the acre," in Peconic Bay. In California, as on the Atlantic coast, it grows more common as you go southward. It is not found north of Cape Cod.

The frond is filiform, growing from a mat of root fibres. on stones and shells, in deep water. It grows in spreading bushy tufts to a height of from three to seven inches. The main stem is as thick as a sparrow's quill at base, thence tapering to the size of a bristle at top. It is irregularly but plentifully branched, especially in the lower part of the frond, the branches spreading out widely in every

* Hypnea, named from *Hypnuma* ; genus of Mosses.

direction, the longest near the bottom. These branches are often branched in the same manner, and sometimes the branchlets also. All the parts are beset, sometimes thickly, sometimes sparingly, with short, horizontal spines one-tenth to one-third of an inch long.

The distinguishing mark of the plant is this: The almost or quite naked extremity of the principal branches is turned back at the ends so as to form a hook, often not unlike a fish-hook in appearance. This must not be mistaken for the twining tendrils borne on the end branches of one variety of *Cysto-clonium purpurascens*. The color is a dark, dull red, with a purplish tinge, which rapidly fades to dirty green and white, when exposed to sunshine or the action of fresh water. It adheres to paper, but not very strongly.

Order.— *RHODYMENIEÆ.*
Genus.— *RHODYMENIA,* * Grev.

RHODYMENIA PALMATA, GREV.

The *palmate* or *hand-shaped Rhodymenia* is so common and so universally known under the common

---

* Rhodymenia = A red membrane.

name of "Dulse" that it seems hardly necessary to
give a particular description of it. As its name says,
it is a red membrane. From a small, hard disk, a
very short, round stem arises for one-fourth of an
inch or so, and then spreads out into a broad, thin,
fan-shaped membrane, three to twelve inches or more
high, destitute alike of midrib and veins. But it is
cleft from top to bottom, or nearly, into many wedge-
shaped segments. The main segments are cleft down
half way or so, giving them also, and the whole plant,
somewhat the appearance of a hand with the fingers
spread out. The margins of the frond are usually
quite entire, but the ends of the "fingers," are cut
in a little way, to show where other divisions would
come.

The plant, however, is variable, sometimes growing
a foot or more high, a narrow leathery strap, fringed
along the sides with leaflets, and surmounted with
several palmately divided segments. It is a perennial,
and the old fronds are generally much thicker than
the young ones. I have some very thin, quite translucent specimens from Sweden. But my British and
Spitsbergen plants are thicker, like our American
forms.

It is of a dark red or wine color. It grows on
rocks, and on the *Fucus*, and on stems of *Laminaria*,

from low-water mark to several fathoms down. It
adheres very imperfectly to paper when dried, unless
allowed to stand for a considerable time before mounting, in fresh water. Both cooked and in a raw state,
it is a common article of food among the peasantry
of the British Isles. In Norway and Sweden, it is
much used as the food of sheep and goats. Mrs.
Bingham reports it at Santa Barbara, common.

Rhodymenia corallina, Grev.

Starting in a cylindrical stem which sometimes is
as long as one-third of the whole plant, it soon
expands into a wide, fan-shaped, many times forking,
rose red frond. The plant is from four to eight inches
high. The lobes, which are generally of a uniform
width in the same plant, vary from one-third to three-
fourths of an inch, in different plants. The margins
of all parts are very entire and smooth, and the ends
nicely rounded. The substance is thin but firm.
It grows in rocky tide pools and in deep water, along
the whole coast of California, very common both north
and south. It is not found on the Atlantic coast.

Genus.— *EUTHORA*, *Ag.*

EUTHORA CRISTATA, AG.

The *crested Euthora* is among our most interesting and beautiful northern plants. Plate XIV. gives
a good reproduction of a typical frond of this species.
In general outline, when spread on paper, it is not
greatly unlike some forms of *Delesseria alata*, from
which it differs, however, by having no veins or
midrib, and by having its end ramifications notched.
In *D. alata* they terminate in rounded points.

The flat fan-shaped frond grows from one to
three inches high, and divides from the base in a
manner between forking and alternate branching.
The main branches also subdivide in the same way.
Sometimes they widen upwards at first, and then
fringe out into narrow branches. Sometimes they
are of the same width throughout, one eighth of an
inch or more, and rapidly divide toward the ends into
minute branches, each of which, under the glass, will
seem to be notched in at the end. It is a full bright
red color.

It is found in great abundance along our whole
coast north of Cape Cod. It has also been dredged
off Block Island. It grows with *Ptilota plumosa*,
and the two *Delesseriæ*, on stones, shells, and other

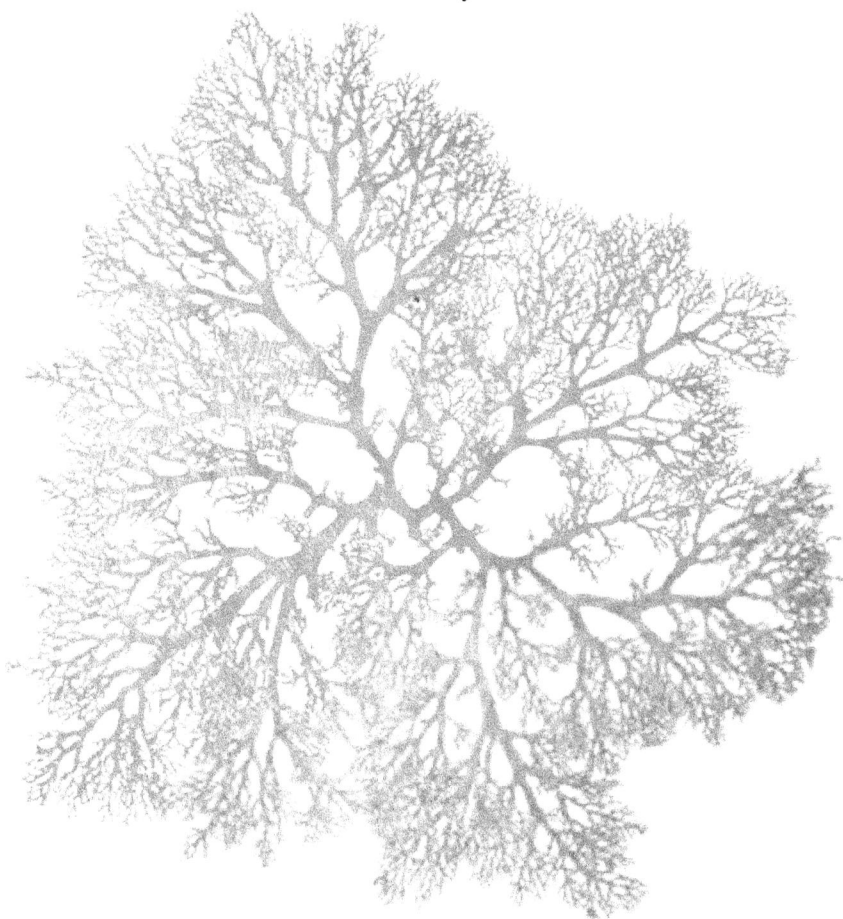

Algæ in deep water. It is to be looked for among the *debris* left upon the strand by the waves. Professor Eaton found it near Eastport, Me. in tide pools, an unusual habitat, I must think. It may be collected throughout the season. It adheres well to paper, and, when carefully laid out makes a beautiful specimen.

Genus.— *PLOCAMIUM,** Lyngb.*

PLOCAMIUM COCCINEUM, LYNGB.

A plant of the *scarlet Plocamium* is well represented in Plate XV. It is one of the most brilliant, beautiful and common of the California Algæ. Few collections of " Sea Mosses " will come from the Pacific coast, which will not contain more or less of them. It grows between tides in pools, and below. Its color is a dark lake red, often faded to a lighter hue. The substance is cartilaginous. The frond is narrow, one-tenth to one-eighth of an inch wide, from three to eight inches high, flattened and branched from the edges, by stout, flattened, alternate branches, some long and some short.

Plants of this species may be easily **and infallibly**

---

* Plocamium = Braided hair.

13

distinguished by the peculiar arrangement of its extreme ramifications. The ultimate ramuli are set on the inner edges of the terminal branchlets, exactly like the teeth of a comb, three or four little awl-shaped teeth in a row upon each branchlet, and the branchlets themselves, set in the same way, upon the edges of the penaltinate branches.

It adheres very well to paper when mounted fresh from the sea, under considerable pressure. It is so common at all seasons, along the whole western coast, that particular localities need not be named.

It is not a little singular, that this species, which is so common on the western shores of both Europe and America, should not be found at all on the eastern coast of America, lying directly between.

Genus.— *STENOGRAMMA, Harv.*

STENOGRAMMA INTERRUPTA,* MONT.

The same remark may be made of this as of the last species; the singularity of its occurrence on the western shores of both continents, and its absence from the intervening east coast of America.

* Stenogramma interrupta = An interrupted mark or line.

PLOCAMIUM COCCINEUM, *Lyngb.*

.

It grows in deep water, on stones and weeds, from a discoid root, with a short stem, which immediately flattens into a thin, wedge-shaped, repeatedly forked membrane, two to eight inches high, widely spreading, the lobes from one-fourth to one-half an inch wide, with parallel sides and rounded apices. The color varies from a pink to a full red.

The fertile fronds may be known by the interrupted or broken line of very dark red fruit vessels, which runs up the middle of the frond and its segments, quite like a midrib. The barren plants have an appearance much like that of *Rhodymenia corallina*, but may usually be distinguished from that species, by their much brighter red color. Fronds bearing asexual fruit are dotted over with irregularly shaped, dark red spots. It is reported on the whole coast of Calfornia, but not very common anywhere.

Genus.—*PIKEA*, *Harv.*

Pikea californica, Harv.

This is a common, coarse, cartilaginous plant, growing between tides at all seasons along the whole California coast. It has a thickish, narrow, flattened frond, one-eighth of an inch wide, three or four

inches high, with a spread of its multitude of branches
all in one plane, in a general fan-shaped outline,
quite as wide as it is high. The flattened branches
spread out widely from the two edges of the main
stem, and divide and sub-divide profusely and irregularly. The only distinguishing point in its outward
appearance is the fact that all the lesser branches
are bordered along both edges by a considerable
number of short, inward-curved, forward-pointing,
spine-like ramuli, of various lengths, from one-tenth
to one-fourth of an inch, short and long mixed in-
discriminately. There seems also to be an utter lack
of system in the branching of the plant. Its color
is a dark red, becoming much darker in drying. It
adheres imperfectly to paper.

Genus.— *FARLOWIA*, *Ag.*

Farlowia compressa, Ag.

This genus, which Prof. Agardh has named in
honor of our countryman, Dr. Farlow, of Harvard
College, who is doing so much fine work in per-
fecting, and disseminating a knowledge of American Algæ, comprises two species, but one of which
I shall undertake to give an account of.

This species is distributed along the whole California coast, is well marked, and, from its outward resemblance to *Pikea*, as well as by its own peculiarities, it will not be difficult to determine.

It has a coarse, tough, leathery frond, narrow, flattened, profusely and irregularly branched from its edges, in a way quite impossible to describe, and yet easy enough to recognize when once seen. It grows to a height of from eight to twelve inches, and has a lateral spread of branches quite equal to that.

Most of the fronds have a well-developed leading stem, though in some it is lost midway in the multitude of branches which spread out each side. Neither stem nor long branches are ever over one-eighth of an inch wide, thickened in the middle, roughened, often toothed along the edges.

The branches and branchlets are all tapered towards the base, and mostly pointed at the top. The ultimate branchlets and ramuli, which are from one-half inch to one inch long, show a decided tendency to bend inward towards one edge like a sabre.

The color is a very dark red, turning almost black in drying. It does not very closely adhere to paper.

The other species, *F. crassa*, I have no specimens or notes of, and so can give no account of
it. It is a northern plant, and may be found from
Santa Cruz northward through Oregon.

Genus. — *CHAMPIA*,* *Ag.*

CHAMPIA PARVULA, HARV.

The *little Champia* is an extremely variable, but
on our southern shores, a very common plant. It
need not be looked for north of Cape Cod. I
have found it in abundance at Southold, L. I., Newport, near the beaches, Martha's Vineyard, Onset
Bay, and at other points. The fronds are fi'iform.
Main stem and branches about the size of a pack-
thread. The living plant, in the water, is apt to
assume a globose appearance, on account of its prolific
and irregular branching. It grows to the height of
from two to six inches. It is softly cartilaginous, and
adheres well to paper. Its distinguishing mark, in
the typical form, is, that both in the water and on
paper, it is regularly and somewhat deeply constricted.
The constrictions vary in length from once to once
and a half times the diameter of the frond. They

---

* Champia = A personal name.

are longest in old parts of the frond, and gradually
shorten towards the ends of the branches, till at last
they appear under the lens, like a string of very small
beads.

In other than the normal forms, these constrictions
are not apparent except to a microscopical examination.
The beginner is advised to put doubtful cases aside,
and wait till a greater familiarity with the species
enables him to be sure of them. I have found the
typical forms to be mostly of a brownish purple color,
darker on paper, while many of the others are of a
decidedly pale green, touched with whitish yellow in
spots, with perhaps here and there brown branches
intermingled. It is a deep water plant, and may be
got through the warm season.

Genus.— *LOMENTARIA, Lyngb.*

This genus is represented by two not very common
species on our eastern coast, but one of which, however, is of sufficient importance to come within the
scope of this book.

LOMENTARIA BAILEYANA, HARV.

This is a very beautiful little plant, growing in
globose tufts, two or three inches high. It is of a

delicate red or pink color, and takes on a variety of interesting forms, one of the most beautiful of which is represented in Plate XI., Figure 2. The normal form is that of a frond as thick as a bristle, forking and branching as it rises, the branches being much constricted at their insertion, and bending in graceful curves towards their extremity. Sometimes the main branches bend over in the long sweep of a semi-circle, as in the plate, and the branchlets springing from the convex side of the arched branch, in their turn bend in the same way, they again being beset externally with arched ramuli.

The normal variety differs from this only in having the parts less bent. But the tapering of both branches and ramuli, to base and apex, is characteristic of every variety. It grows in deep water, four or five fathoms down. It is common south of Cape Cod, and is not found to the northward of that. I found nearly all forms of it at Wood's Holl, in August, and Miss Booth collects it at Peconic Bay, in that month. The *divaricate* form makes an extremely beautiful and graceful picture. It adheres well to paper in drying

Genus.— *RHABDONIA,* Harv.*

RHABDONIA TENERA,† AG.

This genus is represented by one species on each of our American coasts. The one named first is the Atlantic plant. It is found only south of Cape Cod, where it is a very common but somewhat variable plant. In general appearance it is not greatly unlike *Gracilaria multipartita*, differing mainly in color, and in having a cylindrical and not a flattened frond. The stem and branches are somewhat stouter than a wrapping twine.

The plant grows from six to twelve inches high, is very irregularly branched, the branches longest near the bottom of the frond, shorter toward the top, but always attenuated at base and apex. Sometimes the main stem runs through the whole plant, sometimes it is so divided into large branches as to be quite lost sight of. The branches themselves also divide, in a manner between branching and forking, and even the somewhat profuse secondary branches not infrequently have scattered ramuli upon them.

The frond manifests a marked tendency to flatten-

* Rhabdonia = Rod-like.
† Tenera = Tender.

ing, at the point where several branches put out
near together. The fruit is produced on the long
branchlets in hemispherical, wart-like protuberances,
as large as grape seeds.

The normal color is a dark red, which fades on
exposure to the air, and so the plant may come to
have almost any tint, according as it has been for a
longer or shorter time tossed about by the waves,
exposed on the shore, or treated to fresh water in
mounting. It grows upon rocks and stones, several
feet below low-water mark. It is so common everywhere south of Cape Cod, that special localities need
not be named. I have found it everywhere in those
waters.

RHABDONIA COULTERI, HARV.

This species seems to be as common on the California
coast as *R. tenera* is on the Atlantic shores. It differs
from that if I may judge by a somewhat limited suite
of specimens, and from Harvey's description and
figure, by having a more pronounced leading stem,
not branched near the base, and by having all the
branches much shorter in proportion to the length
of the plant, and crowded together towards the top
of the frond.

It grows at low-tide, and below, on rocks, and is
found thrown up upon the beach, somewhat rarely,

from January to March, at San Diego, and all the year around, in great abundance, at Santa Cruz and Santa Barbara.

Order.— *SPONGIOCARPEÆ.*
Genus.— *POLYIDES,* * *Ag.*

POLYIDES ROTUNDUS, AG.

This is the only species in the genus, and the only genus in the order. Agardh names it *P. lumbricalis,* but *rotundus* appears to be the older name. The frond is cylindrical, and rises from a minute disk, at first very slender, then thickens, and at the height of an inch, or an inch and a half, is as large as a knitting-needle, where it widely divides or forks.

In the course of half an inch more, each of the branches forks in the same way; a little further on, all these fork, and again these branchlets, till there are six or eight regular dividings, each successive one being less wide and spreading than the one immediately before it. This gives the plant a fan-shaped outline. The branches all keep their cylindrical form, so that the plant looks stiff and bare, notwithstanding its much branching.

* Polyides = Many-formed.

In color, it is very dark red when fresh, and quite black when dry. It is a perennial, and so may be looked for at all seasons. It grows in deep water. I have taken it at Marblehead and Newport. Mr. Collins reports it in various places about Massachusetts Bay, in the summer and fall, in tide pools. Mrs. Davis gets it at Annisquam in a mill pond. Mrs. Bray finds it washed ashore at Coffin's Beach, Gloucester. All report it common. Miss Booth finds it scarce at Orient. It does not adhere to paper, and is far from being, to the generality, an interesting plant.

Order.—*BATRACHEOSPERMEÆ.*
Genus.—*NEMALION,* Ag.*

NEMALION MULTIFIDUM, AG.

The *many-times-divided Nemalion* is a summer annual, growing attached to the surface of rocks, on the sea bottom, which are uncovered at low tide. It much affects the smooth, rounded surface of the hard, granitic, sea-worn boulders, which lie low down, between tides, all along our New England coast. Where nothing else seems able to make a foot-hold,

* Nemalion = Crop of strings.

or keep its place against the beating of the fierce waves, we often find numbers of these worm-like fronds fastened and flourishing.

At Marblehead, in early June, I have seen these boulders lying clean, smooth, and hard, warming in the sun, when the tide was out, with no trace of vegetation on them. In early July, I have found the young fronds of the *Nemalion* just sprouting up, half an inch high or so. By the middle or last of August, they would be a foot long, full grown, and in perfect fruit. But on visiting the place in October, I have found no trace of them left.

They have ripened, produced the living crop of spores, discharged them into the sea, and so having accomplished their life-function, have vanished again from among living forms.

Where and how the spores pass the intervening months, from October to June, in the midst of the furious waves, and then come back to their native habitat, on the smooth, rounded faces of these bare boulders, there to germinate and grow, and accomplish the circle of their life-history, "is something no fellow can find out;" and it always seemed to me a very wonderful and mysterious thing.

Nemalion multifidum has a cord-like frond as thick as a match, six to twelve inches long. when full grown, very elastic and tough. It divides and sub-divides by regular forkings, the axils being wide and rounded. Sometimes a frond, or a branch, will divide into three or four lobes at the same point, spreading out like the fingers of the hand when widely opened. Again, the forkings will follow each other, in rapid succession, and again, only at long intervals. Usually several, and often quite a bundle of fronds, spring from the same discoid hold-fast upon the rock. The color is dark brown or purple. It shrinks much in drying, and adheres closely to paper. When in fruit, it makes interesting micro-scopical specimens. It is common from Long Island Sound northward. I have found it as plentiful at Newport, as at Marblehead.

Genus — *SCINAIA*, *Bivon.*

Scinaia furcellata, Bivon.

The *forked Scinaia* is not a very common plant, but is worth looking for wherever it is likely to be found, viz.: in our warmer seas, south of Cape Cod, especially at Newport, Gay Head, and Katama, Mass., and in California, where it is said to be quite common. I took several fine plants in Newport in

July. It is a summer annual, of a fine lake-red color, not over four, and usually not over two inches high.

The frond is cylindrical, one-eighth of an inch in diameter, tapering much at the base, sometimes constricted at intervals, and repeatedly and regularly forking as it rises. The frond divides and subdivides six or eight times, and finally ends in little forks, hence its name. All the branches attain the same length, so that the plant is "level-topped," and its outline, when carefully laid out on paper, is almost a perfect semi-circle. It adheres well, and must not be subjected to too much pressure at first. The ultimate branchlets are usually thickened a little. It makes an interesting and sometimes a beautiful specimen. It grows in deep water.

Order.— *GIGARTINEÆ.*
Genus.— *PHYLLOPHORA,* Grev.

The characteristic of the genus is a hard, cylindrical stem, considerably branched, from one to three inches long, and bearing upon the end of the branches a small, wedge-shaped, red leaflet.

* Phyllophora = Leaf-bearing.

PHYLLOPHORA MEMBRANIFOLIA,* AG.

This is the more common species of the two which are natives of our waters. It especially loves the warmer seas, though it is reported as not uncommon on our northern shores. Mrs. Davis collects it at Magnolia, and Mr. Collins at Revere. I found it at Newport and Wood's Holl, in great abundance, especially at the last named place. It grows in deep water on pebbles and rocks. From an expanded disk upon the stone, fifteen or twenty cylindrical fronds sometimes arise in a bunch. At the height of half an inch they begin an irregular branching.

The branches are short and stiff, and stumpy. Some of them soon expand into various sized wedge-shaped leaflets, from one-fourth to three-fourths of an inch long ; others appear merely flattened and then truncated ; others bear the minute lobes of young sprouting leaflets. The typical leaflets are once or twice lobed or forked. The plants grow from one and one-half to six inches high, of a clear red color, and the old ones are often incrusted with parasites, patches of polyzoa or of calcarious Algæ. It is a perennial.

PHYLLOPHORA BRODIÆI, AG.

This is said to be very common in deep water at

---

* Membranifolia = A membraneous leaf.

Halifax, and in northern regions generally. It differs from the last in having a much less branched stem, and a much broader and larger leaflet. Yet this is very variable both in size and form. But the frond is much more simple, and of a somewhat more robust habit than *P. membranifolia.* The leaflet is deeply lobed, but all the segments keep their wedge-shaped outline, and are themselves indented at the top. The color is a clear, strong red. It grows in deep water, and is a perennial. I have never collected it. Mr. Collins finds it occasionally at Nahant, in October, and Mrs. Davis finds it in the fall, on the open beaches, about Gloucester, after a storm. It has been found as a rarity, by Miss Booth, washed ashore at Orient. It has the same geographical range as the other species. Neither of these plants adhere to paper, nor are they especially interesting to the general collector.

Genus.— *GYMNOGONGRUS,* Mart.

This genus is represented by one species on the Atlantic and three on the Pacific coast, in our flora.

GYMNOGONGRUS NORVEGICUS, AG.

The *Norway* species is reported at many places

Gymnogongrus = Naked warts, seed vessels.

14

on our coast, Peaks Island, Me., Beverly and **Nahant**
Mass, and New York, But I do not think it can be
a very common plant, for I have never happened to
find it growing, and none of my correspondents have
seemed to be more fortunate than myself. It grows
in deep water, about two inches high, from a little
disk, by a stem at first cylindrical, twice as thick as
a bristle. In half an inch it forks, sending out a
main branch each way. In half an inch more it
flattens to one-eighth of an inch wide, and forks again
with a wide, rounded axil. Directly these again fork
in the same way, till five or six divisions have been
made, and the ultimate lobes will be one-fourth to
one-half an inch long, standing wide apart, and
rounded at the end. It has a darkish red color on
paper.

GYMNOGONGRUS LEPTOPHYLLUS,* AG.

This plant somewhat resembles the last. Like that,
the frond is flat and narrow, but the stalk is shorter
and not so cylindrical. Starting from a discoid
hold-fast, a small, narrow, flat stem arises, which
either branches at once, or forks at the height of
half an inch, into two widely spreading parts. These
divide and sub-divide, in the same way, two or three

---

* Leptophylus = Thin-leaved.

times. **In a** plant two inches high, none of the parts are over one-tenth of an inch wide, and usually not more than one-sixteenth. The fertile fronds have little hemispherical fruit-vessels scattered over them.

The substance of the frond is thin, but cartilaginous and tough ; the color, a darkish or brownish red. It adheres imperfectly to paper. It grows along the coast northward from Santa Barbara, not very common, on rocks, between tides, at all seasons.

GYMNOGONGRUS GRIFFITHSIÆ, AG.

The color, size, and method of branching of this plant is much the same as that of the last. But it differs from that by not being flat, but quite cylindrical. The frond is not thicker than a bristle. It grows from one and one-half to two and one-half inches high, in tufts, upon rocks, between tides, each frond somewhat regularly forking three or four times. The fruit is held in little, dark-colored, prominent swellings, in the end branches. It has the same geographical range, and the same habitat as the last.

GYMNOGONGRUS LINEARIS, AG.

This is a much larger plant than either of the others, some in my herbarium being not less than

six inches high, and eight inches in the spread of
the frond. The general habit of growth is the same
as that of *G. leptophyllus.*

Rising by a flattened stem, which, two inches
from the base, widely forks, the two parts themselves
fork three or four times. The segments are nowhere
more than one-fifth of an inch wide, and all gradually taper towards the end, the ultimate ones being
long and slender.

The fruit-vessels stand out like hemispherical
warts, one-tenth to one-eighth of an inch in diameter, upon the flat side of the frond. Color of the
plant a dark red; substance, thickish, cartilaginous,
leathery. The general distribution and habitat are
the same as that of the other Pacific species, along
the whole coast of California.

Genus.—*AHNFELTIA,* * Ag.*

AHNFELTIA PLICATA,† FR.

This species is very common from New York
northward, and is also found sparingly at some points
on the west coast. It is extremely easy of identifi-

---

* Ahnfeltia. Named for Ahnfelt, a German botanist.
† Plicata = Folded or doubled up.

cation. **If** you find thrown upon the beach, or growing upon the rocks, between tides, a tangled bunch of black, branched, crooked, very stiff, wire-like sea-weed, half as big as your fist, or larger, the wires as thick as large pins, or knitting-needles, you may be sure it is *A. plicata.*

It is very irregularly and profusely branched, sometimes by widely forking, sometimes four or five branches will grow out close together from the side of the stem, and perpendicular to it ; and the parts spreading and bending by sharp angles in all ways, the plant will be tangled and intricate, beyond description.

Again, it will grow up, and by the upward tendency of the branches, and something like regular forkings, will attain a considerable perpendicular height, six to ten inches, or so, and appear to have some systematic plan of life. These forms, I have collected somewhat abundantly at Newport. But the first-described aspect is by far the most common.

On being exposed on the beach for some time, it will be found faded or bleached perfectly white. It does not adhere to paper, and is altogether as unmanageable a bit of vegetable crookedness and perversity, as one would care to meet. It is too common to require the naming of special localities.

AHNFELTIA GIGARTINOIDES, AG.

This plant is found only on the California coast. It is reported not common at Santa Cruz and quite rare at Santa Barbara. It is a more robust and, by far, less profusely or irregularly branching plant, than the last.

The specimens in my herbarium are six inches high, some of them rising for three inches in a single cylindrical stem, and then forking regularly and evenly in one plane six times, giving sixty-four terminal points to the plant. Others fork fewer times, and less widely, and nearer the bottom of the stem, and then stretch out in long segments two or three inches, before they divide for the second and third time. Like the other, it does not adhere to paper, and its substance is hard and horny when dry. Color, a dark red.

Genus.— *CYSTOCLONIUM,* Kütz.*

CYSTOCLONIUM PURPURASCENS, KÜTZ.

The purple Cystoclonium is a very common, sometimes a provokingly common, coarse, bushy, and

---

* Cystoclonium = Bladdery branches.

generally uninteresting plant. It grows everywhere along our eastern coast, but more plentiful, I think, in our northern waters. At least, my correspondents so report it. It grows between tides, on the rocks, in tide pools, and in deep water.

The main stem runs through the whole plant, thick as a match, somewhat translucent and fleshy, a foot or so high, when full grown. It is irregularly much branched all around, with branches which are themselves branched like the main stem. The ultimate branches are somewhat narrowed at the base, and attenuated into acute points, and sometimes into long, slender, hair-like prolongations at the top.

In variety *cirrhosa*, these attenuated ramuli have the habit of twisting themselves into spirals, like the tendrils of the pea or grape vine, and wind themselves about the branches of neighboring plants, quite after the manner of their more cultivated cousins, the vines. The variety, is perhaps quite as common as the normal form on our shores, and will be likely first to attract the notice of the attentive eye, to the species.

Much trimming will be needed to make the plant presentable on paper. The color varies from a light red brown to a dark purple, or even black, when dry. You will often find that the lesser branches are much

swollen at points, into what appear to be little
"bladders," as the name of the plant mentions.
This is caused by the interior nodules of fruit bulging
the ramulus out at these points. It may be collected
during the whole season. In some places it will
make no inconsiderable part of the mass of smaller
weeds, which are found piled up on the beach.

Genus.—*CALLOPHYLLIS,** Kütz.

One of the marked features, of the marine flora
of California, are the large and brilliant plants of this
genus. None of the red Algæ excel them in brilliancy
of color, and few in size of plant, in spread of frond,
or variety of form. They are common everywhere
on the coast, and grow mostly in deep water.

CALLOPHYLLIS VARIEGATA, AG.

None are more common or more variable than
the plants of this species. It is rightly named. Plate
XVI. shows a common, and what may be considered
a typical form of it. It gives at least the general
method of the division of the frond. And yet many
plants are far removed from this form, by having all
the segments very narrow and long, one-eighth of

* Callophyllis = Beautiful leaf.

an inch wide, and six inches long; or very wide, from an inch to an inch and a quarter broad, and no more than half a foot long.

But the deeply cleft, widely spreading, flat frond, with the segments wedge-shaped, and the extreme ends of all the parts notched in, more or less angularly, are unmistakable marks of the species under all forms. It adheres fairly well to paper. Color, from a darkish to a bright red. The older parts of the plant are thick. The fruit appears in hemispherical warts, scattered over the surface of the frond. Dr. Farlow expressed to me the opinion, that California plants, which have been distributed under the name *C. discigera*, are only extreme forms of *C. variegata*, while those which have been called by collectors *C. ornata*, are really none other than members of the species to be next described, viz. :

CALLOPHYLLIS FURCATA, FARLOW.

Starting from a mere point, where the frond is attached, it widens out till it is from half an inch to an inch wide, and several inches long, and then divides in various ways, mostly by the process of splitting. The clefts are narrow and deep, and some of them run near to the base of the frond; or starting together from the widest part, the clefts run to the end outward, and the segments are arranged like the fingers of the hand, when spread apart somewhat; or the frond may be long and narrow, with an occasional fork.

In every case, except that of the deeply cleft fronds, the lobes are bordered on both edges by a multitude of tongue-shaped leaflets, from one to two inches long, and from one-eighth to one-half an inch wide, much attenuated at base, and with a somewhat rounded point at top. The color is a deep, darkish red. The substance is firm, and in old plants, thick and hard when dry. The fruit, in prominent warts, is scattered over the surface of the frond. The plants in my herbarium range from four to fourteen inches in height. It grows between tides at all seasons, and is not uncommon at Santa Cruz, and other parts of the coast

CALLOPHYLLIS FLABELLULATA, HARV.

This species is more decidedly fan-shaped in outline, and in the division and spread of its main branches, than either of the other species. The principle stem forks, but not widely, and these again fork; then, at a distance of half an inch or so, they divide into half a dozen different segments, each of which repeats the same process, two or three times. The segments are from one-fourth to one-sixteenth of

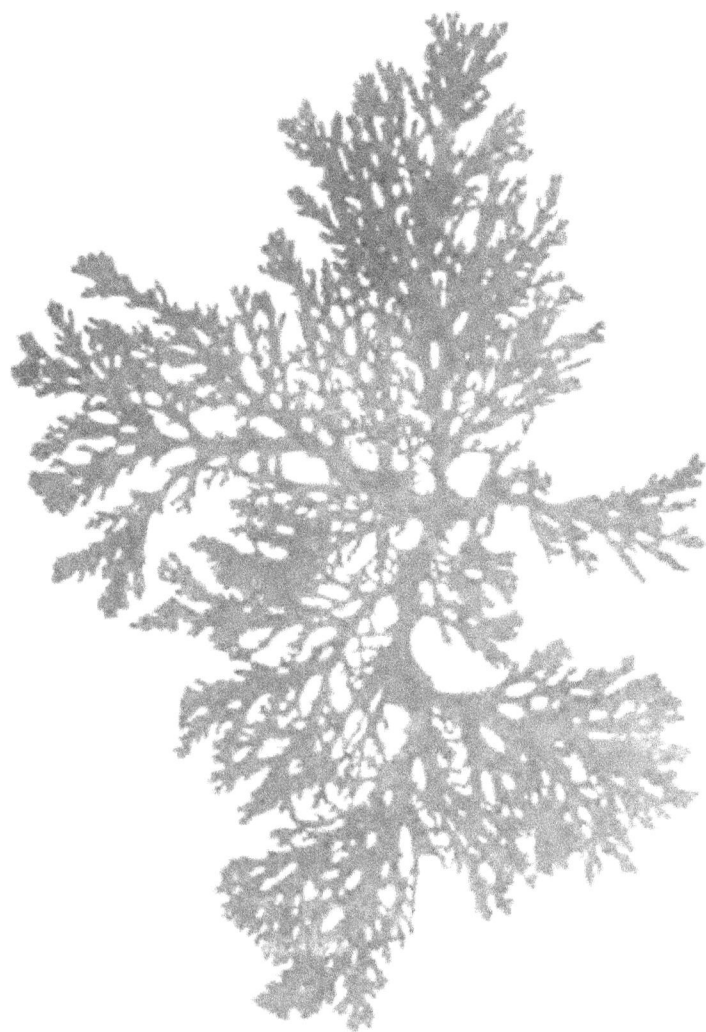

CALLOPHYLLIS VARIEGATA, 45

.

an inch wide, and the extreme ends are notched in, not unlike those of the *Euthora cristata.*

Agardh takes notice that the whole plant resembles some forms of that species. I am not informed whether or not they are commonly found larger than those in my herbarium. These are two inches high, and about three inches wide. The color is a bright rose red, and the substance thin and delicate, adhering well to paper. It is a common plant at all seasons, north of Santa Barbara, and grows between tides and below, on other Algæ.

Genus.— *GIGARTINA,** *Lam.*

This genus, which has several large and showy species on the Pacific coast, and in other parts of the world, has but one, rather humble and insignificant representative, on our eastern shores.

The fronds of the Pacific plants are inclined to be thick, fleshy and bulky; and all the species show, in some form, the presence of the papillose or tuberculose processes, which characterize, and give the genus its name. The plants are of a decidedly gelatinous substance, and one can readily see, that

* Gigartina = Grape stones, referring to fruit-bearing tubercles on the frond.

they might be easily applied to culinary uses in the
same way as the "Irish Moss."

GIGARTINA MAMILLOSA, AG.

This plant grows near low-tide, in Massachusetts
Bay, and northward, upon the rocks, among the
"Irish Moss" or *Chondrus crispus*, which it much
resembles in appearance. It has very much the
same habit of growth, a flattened, leathery, tough
frond, forking from near the base, dividing and sub-
dividing in the same way, broadly and openly. The
segments are more or less wedge-shaped, and have
a tendency to roll their edges inward, toward one
surface, making a channel on that side. It differs
from the *Chondrus*, by having on the inside, or con-
cave side of the frond, a numerous growth of papillose
protuberances. These readily distinguish the plant,
and give it its specific name.

I have collected it in considerable quantities at
Marblehead, and Mrs. Davis and Mrs. Bray find it
among the commonest plants on Cape Ann, as Mr.
Collins does also at Nahant. It is common at Santa
Cruz and northward. The color is a very dark purple,
black and rigid when dry. It does not adhere to
paper.

GIGARTINA RADULA,* AG.

This, and the remaining species of this genus, are exclusively natives of the Pacific coast. This is the largest and most pretentious species of the genus. It has a large, flat, thick, dark, livid red frond, which takes on in different plants quite a variety of forms and outlines. But in the main, it is simple, or if divided, then only by the presence of one or two clefts of greater or less depth.

It puts out no branches or leaflets, but is more or less thickly peppered over with warty protuberances, which seen along the edges of the frond in profile, appear to be mostly minute globes, a half or a quarter as large as a pin head, set upon short stalks.

The frond itself rises from a short, flattened stem, from which it more or less rapidly widens to a breadth of several inches, then, in the simpler forms, rounds off, usually very bluntly, at the top. The largest specimen in my herbarium is fourteen inches long, and six inches wide in the middle, tapering more rapidly and acutely to the top than to the bottom. But another specimen, ten inches long, and four and a half broad, tapers quite acutely to the base, and is very broad and blunt at top, even cut in, heart-shaped.

* Radula = A scraper.

I have seen much larger plants than either of these. The variety *exasperata*, grows two or three feet long, and six to ten inches wide. But the heavy, thick, mostly simple, flat frond will serve to distinguish this from either of the other species. My California correspondents all report it very common from San Diego to Santa Cruz, growing between tides, on rocks the year around, or below tide, and in the sluice ways. It is truly a noble plant, and with its livid red color must be a striking feature, rising and falling in the green waters.

<h3 style="text-align:center">GIGARTINA SPINOSA,* KÜTZ.</h3>

This resembles the last species only in its thick, leathery substance, and its roughened, spiney surface. The protuberances are pointed, and not rounded at the end, as in *G. radula*, and they often attain considerable length.

The form of the frond is extremely variable. Sometimes it rises from a cylindrical stem, flattens broadly, and then divides, as the hand divides into fingers. Again, it keeps its main frond entire, and simple, tapering gradually and gracefully to base and apex, and throws out from each edge a multitude of long, narrow leaflets, pointed above and below. These are some-

---

* Spinosa = Thorny.

times simple, and sometimes forked, from one to three inches long, and from one-eighth to one-third of an inch wide.

Both the main frond and the leaflets are covered with a profusion of the stout spinose, or papillose processes peculiar to the genus. Color, a dark red, brown, or purple. It grows from six to twelve inches high, upon the rocks, between tides, and below, at all seasons. Dr. Dimmick and Mrs. Bingham report it very common at Santa Barbara, upon the rocks near shore. But Mr. Cleveland at San Diego, and Dr. Anderson at Santa Cruz, find it not so plentiful as the last, or the next species.

GIGARTINA MICROPHYLLA,* HARV.

The most characteristic difference between this plant and the two preceding species, is its much lighter and thinner frond, and its slenderer, spore-bearing spines. It rises from a disk by a flattened short stem, which more or less rapidly expands into a wide, thin, flat frond. This remains simple or else divides into two or three segments, each of which tapers into a long, slender, pointed apex. This attenuation of the plant at the top, seems to be character-istic of the species. It is thickly covered with the

Microphylla = Small-leaved.

long slender spines, and often bears a few small,
thin leaflets along its edges. It grows to the height
of twelve or sixteen inches or more, and is an inch
or an inch and a half wide. The color is a deep,
brownish red. It is abundant along the whole California coast. It may be found near the wharf, at
Santa Barbara, and at the beach, and mussel beds,
at La Jolla, San Diego.

A plant, which the botanists have insisted upon
calling a variety of this, var. *horrida*, but which differs
from it in all respects, quite as much as *G. spinosa*
does, is very common along the whole coast. It is
a much smaller plant, thicker, and darker colored,
and vastly more profusely and irregularly divided, and
branched, than the typical form. It is literally clothed
in almost every part, with long, closely set, simple
or branched spines. Its appearance well entitles it
to the cognomen "horrid." It is present in considerable
numbers, in almost every gathering of California
"Sea Mosses" which one gets. Why it is not worthy
of a regular specific "local habitation and a name,"
is more than appears clear to me.

GIGARTINA CANALICULATA, HARV.

This, also, is a very common species on the California coast, but quite unlike any other representa-

tive of the genus found there. It rises from a few matted fibres in a narrow, flattened stem, one-tenth of an inch wide, whose edges are slightly turned upon one side, making a channel on that side, and leaving the other slightly convex. It is bare for an inch or more, and then forks or irregularly branches from its two edges. The opposite branches divide and sub-divide once or twice, after a more or less pinnate fashion. The ultimate ramuli, which are minute spines, often bear the fruit in swollen and rounded vessels, developed in their middle in such a way as frequently to turn the end of the spine down at right angles to its general axis, so as to make the whole bear a striking resemblance to a minute bird's head, bill and all.

It grows in dense tufts, from two to four inches high, in tide pools, and on the rocks between tides, all the season through. Dr. Dimmick collects it at Castle Point, Santa Barbara, but it may be looked for, I suppose, in favorable localities everywhere. The younger parts of the plant adhere well to paper.

Genus.— *CHONDRUS,*[*] *Lam.*

CHONDRUS CRISPUS,[†] LYNGB.

This is the famous " Irish Moss " of commerce. It is collected in large quantities on our eastern coast, exposed to the sun to dry and bleach, and then sold to the grocer for his customers to make *blanc mange* of. It grows very common upon the rocks between tides, and a little below, and is as variable a plant as it is common. It is so well known in the East that it hardly need a special description. For others, I may, perhaps, venture to append a brief account.

The fronds are from three to six inches high; thick, tough and leathery. At first, it is a flattened stem; this, at the height of an inch or more, when it is from one-eighth to one-half an inch broad, forks widely. Thence, at varying distances, the parts divide and sub-divide, in the same way five or six times. The frond exhibits all the possible variations between the long and narrow, and the short and wide, and all shades of color, between an olive green and a very dark purple, or jet black.

The purple and other dark shades are apt to be

---

* Chondrus ✻ Cartilage.

† Crispus ✻ Curled.

sheeny, or iridescent, in the water, and are sometimes among the most beautiful plants to be found
growing in the tide pools, especially when the sun
shines upon them. It turns much darker, and does
not adhere to paper, in drying. Its geographical
range is from the Carolinas north, on the east coast.
It is not found on the Pacific side of the continent, though two other species of the genus, which
I have not thought it best to give an account of,
do occur there, viz. : *C. canaliculatus* and *C. affinis*,
the latter of which, Dr. Farlow thinks, may be a
variety of the former.

Genus. —*IRIDÆA,** *Bory.*

IRIDÆA LAMINARIOIDES, BORY.

This species sufficiently characterizes the genus.
It has a large, wide, thick, membraneous frond, aris-
ing from a stalk two inches long, which is at first
cylindrical and then flattened. The frond is usually
simple, though sometimes lobed ; from one to two
feet long and from one to three inches wide, smooth
when barren, warty when bearing the true fruit, and

---

* Iridæa = Many colors.

thickly dotted over, when bearing tetraspores, **with** small, colored, raised spots.

Dried, the plant is stiff, substantial, and tough, and of a very dark red color. It is among the commonest of plants at Santa Cruz, at all seasons, near low-tide mark on the rocks, and in tide pools. It is very scarce at Santa Barbara, growing on small rocks near low-tide, and is altogether absent at San Diego. No representative of the genus is found on our eastern shores.

Order.— *CRYPTONEMIEÆ.*

Genus.— *PRIONITIS,** Ag.*

This is a very common form on the whole of the west coast. The genus is characterized when dry, by a thickish, hard, smooth, leathery, flat frond, of a dark red-brown color.

PRIONITIS LANCEOLATA, HARV.

The specific name refers to the lance-shaped leaflets, which are found upon the edges of its branches. The plant has a narrow, flattened frond, one-tenth of an inch wide, which sparingly forks, or branches from its two edges, in a very irregular,

---

* Prionitis = A little saw.

straggling manner, usually with long distances between the divisions. Although it is an extremely variable plant, it is not difficult to recognize, when once known, as it contrives, in some way, to show its specific peculiarity, viz.: the putting out of minute lance-shaped leaflets, along the edges of the long, ultimate branchlets, which always stand out perpendicularly to the axis of the branch. These are very much constricted at the base, but rounded more or less at the top, and are from one-sixteenth to one-half an inch in length. The plant attains, in full growth, a height of ten inches or more.

Mr. Cleveland finds it, from October to May, washed upon the shore from deep water, at San Diego. At Santa Barbara, it is found in the same situation, also growing on the rocks near shore. Dr. Anderson finds it on shelving rocks and in tide pools, all the year, at Santa Cruz. It is extremely common everywhere.

PRIONITIS ANDERSONII, EATON.

This is a much larger plant than the last. It is common at Santa Cruz, but somewhat rare on other parts of the coast. It was named by Prof. Eaton, for that most industrious and zealous Algologist, Dr. Anderson, of Santa Cruz. The plants are a foot or more **high, and** usually consist of a main frond, which is flat, thick, and of a dark red color, tapering to a point above and below, with a marked tendency to bend toward one edge like a sabre. This may be the whole of the plant, and then the frond will measure a foot in length, and an inch in width, at the widest part.

Commonly, however, this is but the central part of a large and widely-spreading plant, the secondary fronds, branching from the sides of the main frond. Sometimes, this may. be comparatively small, no more than two inches long, and three-tenths of an inch wide, and throw out on each edge a considerable number of long, flat, tapering, sabre-shaped frondlets, perhaps, a foot or more long. Again, the main stem may be three times as large every way, and the branches no more than four or five inches. So they vary in relative size and proportion. The plants of this species are usually of a deep red, wine color. They do not adhere to paper.

Genus.— *SARCOPHYLLIS, Ag.*

SARCOPHYLLIS CALIFORNICA.

This and another species, *S. edulis*, Agardh takes

from the old genus, *Schizymenia*, to make this new genus of.

It has no stalk, but expands upwards into the wedge-shaped base of the broad, thickish membrane. The one before me, kindly lent by Prof. Eaton, is not more than five inches long, but is quite two inches wide at its widest part, tapering to a rounded point at the top. The membrane is simple, but more or less torn. The color is a dark purple, darker in drying.

It is not very common at Santa Cruz, growing on rocks and weeds, on rocky beaches. It is not else-where reported in California, and it does not occur at all on our eastern coast, though its generic con-gener, *S. edulis*, is common enough on the west coast of Europe.

Genus.— *GRATELOUPIA,* Ag.*

GRATELOUPIA CUTLERIÆ, KÜTZ.

This is a large, coarse, flat, extremely variable plant, quite common on the California coast, except in the extreme south, where Mr. Cleveland sets it

* Grateloupia. Named for Dr. Grateloup, a French Algologist.

down as a rarity. It often attains the height of two or three feet. Sometimes the frond will be perfectly simple, an inch wide. and two feet long, tapering to a narrow base and apex; sometimes a foot high and three or four inches wide; smooth and blunt at top; colored so as to closely resemble a frond of *Iridæa laminarioides*, from which then, it is possible to distinguish it only by a microscopical dissection, of the structure of the plant. Again, it will be deeply cleft into many lobes from near the bottom to the top; and, at other times, it will put out a series of leaflets from both edges; or it will combine both these departures from simplicity in one plant; or it will throw out from the truncated top of a long, wide, simple frond, a number of long, narrow frondlets, much attenuated at each end.

The color is a reddish brown, changing by fading to various shades of brown and purple, and even to a dull green, or dirty white. Sometimes all these colors will be found in the same frond. It grows in deep water, plentiful in the north. Dr. Dimmick finds it very common near the light-house, at Santa Barbara. It may be looked for at all seasons.

Order.— *DUMONTIEÆ.*

Genus.— *HALOSACCION,* Kütz.*

HALOSACCION RAMENTACEUM,† AG.

This is truly an Arctic plant, growing only in northern waters, but there sufficiently plentiful. So far as I know, it has not been found south of Gloucester. Mrs. Davis finds it in deep tide pools, from April to August, at Brace's Cove, Gloucester; and Mrs. Bray on rocks, in tide pools, plentiful at Bass Rocks, Gloucester. Harvey figures it as a plant twelve to fourteen inches high, when full grown; with a pronounced leading stem as thick as a crow's quill at the middle, much attenuated at the base, and somewhat so at the top; clothed on all sides above the middle with an abundance of branches, half as large as the main stem, from one to three inches long, mostly simple, but sometimes branched, and always attenuated at base and apex. Both stem and branches are hollow.

My American plants are of a decided red color; but I have Spitzbergen plants, from Prof. Kjellman, of Sweden, which are of a dull purple color,

---

* Halosaccion = Sea-bag.

† Ramentaceum = Branched.

and differ from Dr. Harvey's figure in the much
greater length of their branches. Prof. Eaton describes a variety which he calls *gladiatum*, found in
abundance at Eastport, Maine, which differs much
from the normal form. It is flattened, wide, near
one inch in the middle, but sword-shaped and attenuated at both ends; sometimes simple, and sometimes branched on the edges. Some specimens in
my herbarium show tendencies toward that form. It
is a variable but not uninteresting plant, and collectors
along the coast of Maine, and the Provinces, will not
fail to find it in plenty, on the rocks, near low-tide.

Order.— *SPYRIDIEÆ.*
Genus.— *SPYRIDIA,** Harv.*

SPYRIDIA FILAMENTOSA, HARV.

This plant is an inhabitant of the warmer seas.
It is found common only on our southern shores.
I know of no well authenticated case of its having
been found north of Cape Cod. But south of the
Cape it certainly is as common as almost any plant.
I certainly found it in abundance at Newport, from

* Spyridia = A small basket, referring to the fruit.

July to October, and in Providence River, in August. Miss Booth found it not uncommon at Peconic Bay, and other points about the east end of Long Island. It is also reported by Harvey, at various places in our southern waters, as far as Key West.

The frond is filiform, not usually thicker than a bristle, from three to six inches or more high, generally much and irregularly branched, the branches spreading widely, and being themselves divided and sub-divided into a wealth of lesser ramifications. The branchlets, when young, are visibly articulate ; and all of the smaller branches, and often all the branches, are clothed throughout with a light growth of very delicate, hair-like filaments, not much over one-tenth of an inch long. These are plainly visible to the naked eye, and give the name, and characteristic mark, of the species. The color is a purplish red, but the hue may change by fading through all shades to a pale green or yellow. It grows below tide marks, a fathom or two, and so must be looked for, among the floating burden of the sea. It adheres fairly to paper, and with its fine and gracefully disposed branches, and its soft haze of fairy filaments, bordering all, it makes a very pretty specimen.

Order.— *CERAMIEÆ.*

Genus.— *MICROCLADIA,* **Grev.**

MICROCLADIA COULTERI, HARV.

Probably very few people collect "Sea Mosses on the Pacific coast, who do not get plenty of this species with every gathering. No package of dried Algæ, or fasciculus of mounted specimens, comes from that coast, to the botanists or lovers of Algæ in the east, which does not contain some of these interesting and beautiful plants.

It has a cylindrical or slightly flattened stem, twice as thick as a bristle which runs fully through the plant, and sends out branches from its two edges, in one or the other, or both of the following ways, viz.: The regularly alternate branches are set on the two sides, at an almost perfectly uniform distance, and rise at the same angle from the main stem, so that they "lay out" quite parallel. Near the base, the branches are short, but gradually become longer towards the middle of the frond, then shorten again, towards the apex, so as to give the whole plant a quite perfect "lanceolate" outline. Or, again, the plant will throw out several long branches from each

* Microcladia = Minute branches.

side, near the base, and each of these, together with the main stem in its upper part, will develop the typical outline just now described.

It remains to be said that the primary branches themselves, branch in the same manner, by short, alternate, secondary branches, and these, again, divide up in the same regular way, the ultimate ramuli at the end being invariably incurved, and growing shorter and shorter to the end of the branch. This regular habit of branching, the graceful outline of the plant, and the many shades of red and delicate pink which it assumes, make it a great favorite with collectors.

It does not adhere very well to paper, and on that account is all the more easily detached, and woven into those beautiful "Sea Moss" pictures, which some of the fair admirers of these plants are fond of making. With them, this plant becomes a great favorite. Its fine and delicate ramifications, and its great faithfulness in retaining its normal shape, when once pressed and dried, make it very serviceable for such uses.

It attains a height of six or eight inches. It may be found at all seasons in great abundance, on the rocky beaches, between tides and below, upon rocks and other Algæ, especially upon *Gigartina radula.*

Microcladia Californica, Farlow.

In general form and substance, this very much resembles the last species, but differs a little in the disposition of the ultimate ramuli. But a perfectly unmistakable mark may be found in the position of the fruit. And it would not be exactly safe to call any specimen *M. Californica*, which does not demonstrate its identity by having fruit.

In *M. Coulteri*, the fruit is borne on the inside of the ultimate ramulus, and is surrounded by a little whorl, of incurved, short, spine-like processes, which partly inclose it. In *M. Californica*, the fruit is borne on the outside of the ramulus, and is bare, and destitute of this inclosing whorl. The species is not as common as the last, but is found growing in the same situations along with that.

Microcladia Borealis, Rupr.

Our artist has given such a good picture of this beautiful plant, in Fig. 2, Plate VII, that it cannot be necessary to enter into a detailed verbal description of it. There is nothing in the waters of the Northern Pacific that can possibly be mistaken for it.

It will be observed that the very graceful outline of the plant, is obtained by carrying out, in detail, a perfectly uniform and very simple method of branch-

ing, viz. : putting every secondary branch upon the inside of its primary, and bending the primary outward and backward. This plant could hardly fail to give a fruitful hint, for a decorative design, to any artistic mind.

It is found only in the northern waters of the Pacific, as its name implies. But it is common at Santa Cruz, in tide pools, at all seasons. It is of a very dark brown color, often almost black. It does not very perfectly adhere to paper, and so like its "next of kin," *M. Coulteri*, it becomes a very useful plant in working out beautiful "Sea Moss" designs.

Genus.— *CERAMIUM.** *Ag.*

This genus furnishes several of our most common and most beautiful "Sea Mosses." There are plenty of good reasons for all being favorites with collectors. The distinguishing characteristics of the genus are either or both of the following, viz. : 1. The tendency of the tops of the branches to bend in towards each other, the last fork being quite incurved and hooked, like two minute fish-hooks, turned point to point. 2. The variegation of the stems and branches, as seen with a good pair of eyes, or

* Ceramium = A pitcher, referring to fruit.

under a pocket lens, by alternate bands of lighter
and darker color, sometimes white and black, sometimes white and red, and sometimes two shades of
red. This characteristic never fails, except sometimes
in the older parts of very robust specimens of *C.
rubrum.*

CERAMIUM RUBRUM,* AG.

This plant is common, not only throughout our
entire eastern and western coasts, but in almost every
sea upon the globe. I doubt if there is another so
thorough-going cosmopolite, in the whole marine flora
of the world.

It grows upon everything, rocks, and stones, and
shells, and almost all sorts of sea plants. This abil-
ity to be on a good footing with every kind of companionship, and to feel at home wherever it can find
a place to stand, and sprout, and grow, will account,
perhaps, for its universal presence and its wide distribution.

It grows in pools, between tides, and in deep
water. It is extremely variable in appearance, and
will sometimes almost "deceive the very elect," into
believing they have found some other species. It
grows from two to ten inches high, thicker than a

* Rubrum = Red.

bristle in the larger parts, often, indeed, as stout as wrapping-twine, and always has a coarse appearance.

It branches mostly by forking, the lower divisions distant, the upper ones nearer and nearer together, sometimes narrow, and sometimes widely spreading. The segments attenuate as they divide. The apices are either slightly incurved or quite hooked. The variegated bands are less plainly marked in this, than in either of the other species to be described, and rarely appear as other than light, or dark shades, of the prevailing red.

The microscopist will find the plant covered throughout with a coating or "bark" of small cells. In the other species to be described, this coating is not continuous, but extends only as rings, of a red or dark color, about the nodes or joints of the frond. This is a sure guide to it in all the many forms which the species will assume.

To the collector. who depends upon his eyes and his pocket lens, the deep, full red color, which, indeed, may be faded out by exposure, the general appearance of coarseness, combined with the incurved or hooked apices, will be a sufficiently safe ground for saying that his plant, as he pulls it from the water, is *C. rubrum.*

CERAMIUM DESLONGCHAMPSII, CH.

This species Harvey describes as *C. Hooperi,* in honor of his friend, Mr. J. Hooper, of Brooklyn, N. Y., an enthusiastic and intelligent Algologist, who with Professor Bailey and others, as I have already mentioned in the "Introdction," did much in that time, to help forward Harvey's study of our plants. They all find ample acknowledgement in the pages of the "Nereis."

But it is conceded now that this is no new species, but an old and not uncommon one, on the shores of Europe. It is common along our northern coast, north of Nahant. I found it in plenty at Marblehead, and Mr. Collins at Nahant on the sides of perpendicular rocks, overhung with *Fuci.* Mrs. Davis collects it on rocks in tide pools at Gloucester. Professor Verrill found it on the piles of the wharf at Eastport, and Mr. Prudden at Grand Manan. It grows from two to four or five inches high, from a mass of creeping filaments. The fronds are not much coarser than human hair, and divide throughout by true but not very wide, forkings. The apices are attenuated, sharply pointed, and but slightly incurved or bent, mostly straight or awl-shaped.

Under a lens the markings or variegated bands

are clearly seen. The dark ones keep the uniform proportion of being almost exactly as long as broad, or quite square in every part of the frond. The white bands vary very much in length, and are longest in the old parts of the plant, and gradually shorten toward the apices. The color is a *dark purple*, which sometimes is given out in pressing and drying, so as to stain the paper red or purple. It may be looked for, all the collecting season through, on the sides of perpendicular rocks near low-tide mark.

CERAMIUM STRICTUM,* HARV.

This is probably the species which Harvey describes in the "Nereis." under the name of *C. diaphanum*. Nothing is more common than it and the next species, except it be *C. rubrum*, all along our southern shores. The plant grows in tufts, from two to four inches high, as fine as hair, and divides or branches, by narrowish forks, more and more close, towards the extremity of the frond.

The variegated appearance of the frond is plainly visible to the naked eye. The dark red or purplish bands, are relatively very short, especially toward the base of the plant, where the white interstices are

* Strictum = Drawn together, close, tight.

three or four times longer than broad. Toward the
apices these shorten, till they are not much longer
than the colored bands. The apices are sometimes
only incurved, but more frequently hooked.

It may always be distinguished from the next,
with which only is it likely to be confounded, by
its somewhat greater length; its narrower forking;
its decided tinge of purple in the prevailing
red, of the dry plant; and the fact that the
fronds of a tuft appear to be of a considerably
different length, so that the outline of the mounted tuft
will be decidedly uneven and jagged. I collected it
in abundance at Newport and Wood's Holl, in the
summer and fall. I have never found it in Massachusetts Bay. But Mr. Collins reports it as not uncommon
in the warm waters, on the muddy bottom of Mystic
river marshes, about Boston. And Mrs. Davis collects
it in the river, at Little Good Harbor, Gloucester.

CERAMIUM FASTIGIATUM,* HARV.

This I consider our most beautiful *Cramium*. It
is very common at all points, where I have visited
the south shore of New England and New York,
especially at Newport, where I took hundreds of
splendid plants. It grows on *Zostera*, *Chondrus*

---

* Fastigiatum = Sharp pointed.

CERAMIUM FASTIGIATUM, *Harv.*

CALLITHAMNION CORYMBOSUM. *A*

crispus, and other Algæ, in pools, or below tide. Its most usual form is that of a dense globose tuft, from one to two inches in diameter, of a brilliant red color.

It is very easily seen and caught, as it comes floating in upon the waves. Examined particularly, it will be found finer than human hair, of much the same thickness throughout, branched by wide forkings, the forks coming nearer and nearer together, toward‹ the end of the frond, see Plate XIX., Fig. 2.

The beautiful pink bands are, relatively to the colorless interstices, very short. They are, in fact, shorter than the diameter of the frond, so that under the lens, they appear to be rectangular patches of color, longer crosswise than lengthwise of the frond. The white spaces between, shorten as we proceed from the base to the top of the frond, thus bringing the colored bands closer and closer together.

The filaments in the tufts are of the same general length, as are also their several divisions. This makes the tuft level-topped, and produces that globose appearance which is so characteristic of the species. It also causes that constant tendency of the plant, when mounted on paper, to display its terminal branchlets in some segment of a circle. This difference in outline, the shorter and more uniform length of the frond, and the more brilliant pink color, with no admixture of purple, easily distinguish this species from the last.

C. arachnoideum, which Harvey figures and describes, Table XXXIII. B., of the " Nereis," Dr. Farlow thinks may be a variety of *C. fastigiatum*, but is in doubt. He declares, on the authority of Agardh, that it is not the same as the species of that name in the European flora. I took it in unlimited quantities, in the little harbor at Wood's Holl, the last of October. It is, indeed, a very beautiful and interesting plant.

Genus.— *PTILOTA,* * Ag.

The plants of this genns, which contains two eastern and three western species, are characterized by their cartilaginous, flattened, narrow, pinnately branched, feathery or fern-like fronds. The two eastern species may be easily distinguished by the relative fineness and the place of growth of the two plants ; the three western, by certain marked peculiarities of appearance and ramification.

PTILOTA PLUMOSA, AG.

The var. *serrata* of this genus is a very common

* Ptilota = pinnated, furnished with plumes.

plant in deep water, all along our coast, north of Boston. It grows, attached to rocks and stones, in the bottom of the sea, and to the stems and roots of *Laminaria.* It will be found in great abundance on all open beaches where the waves have deposited it, brought up from the depths.

The frond is three to six inches in extent, one-sixteenth to one-eighth of an inch wide, flattened, tough, cartilaginous, irregularly, pinnately branched from the edges, branches likewise flattened and branched from their edges, all in one plane. Plate XVII., is an exact copy of a specimen in my herbarium, and very fairly represents the beauties of this plant, as well as the pinnate method of branching, common to the genus.

The peculiarity of the species is the dissimilarity of the opposite pinnæ on the ultimate branches. From the branches there will spring forth on one side a beautiful little plume or pinna, while exactly opposite to it, will be a short, curved, undivided spine-like process, somewhat thickened, and often toothed on the outer edge; all the ultimate divisions stand out almost at right angles to the branches. The color is red. A perennial, perfect in summer; adheres, but rather imperfectly to paper. It need not be looked for south of Cape Cod, but it is found on the coast of California, and in the north Pacific very common.

PTILOTA ELEGANS, BONNEM.

This is a much more delicate plant than the last, narrower, thinner and of a darker color. It is common from New York northward. It may be found almost always growing upon the perpendicular sides of cliffs, under the overhanging " Rockweed," near low-water mark. That is the only situation in which I have ever seen it growing. But I have collected it in no little abundance about the beach, at Newport, in the summer and fall, among the mass of sea weed left by the waves. There, it must grow in deep water.

The fronds are nearly cylindrical, but branch like the last, from opposite sides in one plane, decompound pinnate, the pinnæ and pinnulæ opposite and *alike*, though, I think, in most cases one of them is apt to be much smaller than the other. The large and small parts alternate, so that the symmetry of the frond is maintained. Often the smaller pinnule is suppressed altogether, and the branching will thence seem to be alternate.

The ultimate ramuli are composed of a single row of square or oblong cells. This is a fine, delicate and beautiful plant. It adheres well to paper. The

young plumules make beautiful microscopical specimens, if mounted in some fluid which does not shrink
the cells. The beauty, as well as the interest of
the specimen, will be enhanced if the plant bears
upon the tips of its plumules, the tetrasporic fruit.
The color is a darkish purple, more or less red in
the younger parts.

PTILOTA DENSA, AG.

This and the two following species belong to
California. The frond is compressed, one-eighth of
an inch wide, thick, cartilaginous, from three to
twelve inches high. The leading stem bears along
its edges stout branches, which are either simple or
branched, on the same plan as the main stem. The
axils of the primary branches make an angle of about
45°. The ultimate pinnæ, which clothe the edges
of the whole plant, are closely set, making a dense
border to the frond, of very uniform length, one-
tenth of an inch or so, opposite, and *very unlike*.
The one is stout, undivided, incurved, sharply toothed
on the outside ; the other opposite, slender, much
shorter, pinnately and widely divided. The latter is
seen to lie almost hidden out of sight, under the over-
arching pinnule which grows next below it. For,
it will be observed that, the two forms alternate with

each other quite regularly, on both sides of the plant.

This species is a much more robust plant than either of the other California *Ptilotæ*, thicker and denser, every way in appearance. That fact will commonly serve to distinguish it from them. But there are other points which help the discrimination, viz. : the *ultimate simple* pinnule of this species is sabre-shaped, arched or incurved, and toothed on the outer edge only ; while theirs is relatively smaller, straighter slenderer, more club-shaped, and in *Pt. hypnoides*, not toothed, while in *Pt. asplenoides* it is commonly toothed on both sides.

It grows in deep water. Mr. Cleveland gets it from January to April, not very common at San Diego. Dr. Anderson reports it not very common, on the beach, at Santa Cruz, all the year round. Mrs. Bingham says it is rare at Santa Barbara ; she finds it, in February, washed ashore from deep water.

PTILOTA HYPNOIDES, HARV.

I have plants of this species quite two feet long. It greatly resembles *Pt. densa* in its general habit of growth. It has a prominent leading stem, flattened, branching irregularly along either edge, with long, widely spreading branches. These also are beset by shorter secondary branches in the same manner, so that the whole plant lies in one plane. The secondary branches bear the pinnæ. These are opposite and *unlike.*

They consist of a prominent, somewhat bent, thick, club-shaped, obtuse, untoothed ramulus, one-tenth of an inch long, set opposite a smaller pinnately divided pinnule. The smaller divisions of this pinnule seem to be in form like the large, undivided ramulus, which is placed opposite to it on the plant. The divided pinnules seem to be quite insignificant, and are often almost suppressed between the stout, self asserting ramuli by their side.

It does not adhere to paper very well. In color it is a reddish purple, fading to green or a dirty white, older parts often almost black. Mr. Cleveland says it is a rare plant at San Diego, cast up from deep water, from November to April. Mrs. Bingham reports it not very common at Santa Barbara, in May, and June. But Dr. Anderson finds it common at Santa Cruz. It evidently loves a northern climate.

PTILOTA ASPLENOIDES, AG.

This is a still more distinctly northern plant than the last. It is reported in California, only at Santa Cruz, and there as scarce. It is a very much slenderer plant than the last, though growing to the height of eighteen inches. The frond is compressed or flat; one-tenth of an inch wide, of uniform breadth, with a leading stem, and branches and pinnæ on both edges; the axils of primary and secondary branches narrow, while the pinnæ are set almost at right angles to the axis of the branch. They are opposite and unlike.

The larger pinna or ramulus is undivided, one-eighth of an inch long, or less, deeply and sharply toothed on both edges, widened in the middle, and pointed at both ends. The opposite pinna is either reduced to a minute spine or pinnately divided, but always much less prominent than the ramulus, which sets opposite to it. The color is a light or reddish brown. It does not adhere to paper.

Genus — *GLOIOSIPHONIA*,* *Carm.*

GLOIOSIPHONIA CAPILLARIS, CARM.

This is often spoken of as a rare plant, but I have found it so common in the rock pools about Marblehead, that I can hardly think of it as rare

---

* Gloiosiphonia = A viscid tube.

or even scarce. It is said to be found in Long
Island Sound, but where, or in what part of it, or
the adjacent waters, I am not able to say. It
more properly belongs to our northern waters, and
from various points there, it is reported. Mr. Collins
finds it at Revere, in tide pools, in June. Mrs. Bray
finds it in deep water at Magnolia; and Mrs. Davis
collects it from May to July, at the same place, on
rocks partly covered by sand.

It grows six or eight inches high; the main stem
cylindrical, as large as wrapping-twine; sometimes
solitary, but commonly in tufts. It is much constricted at the base, and attenuated at the top, as
are also all the branches and the ramuli. It has a
leading stem, which, at the height of an inch or
more from the base, begins to be clothed with short,
widely-spreading, almost horizontal branches. In a
plant six inches high, some of them exceed an
inch in length. They are inserted all around, and
somewhat evenly distributed along the main stem.
They branch in the same way, and the secondary
branches are also beset with ramuli, arranged on the
same plan. All the parts are much constricted at
the base, and attenuated at the top.

The substance of the frond is soft, or tender
and juicy, and a little elastic, shrinking much in

drying. It adheres firmly to paper, and should not be subject to much pressure, at first, in drying. The color of the younger plants is a brilliant carmine; older ones, darker. It should be looked for early in the season, though I have collected it to the end of August.

Genus.— *GRIFFITHSIA,** Ag.*

GRIFFITHSIA BORNETIANA, FARLOW.

This is the only representative yet found on our eastern shores of this large and brilliant genus. It is called *G. corallina*, var. *globifera*, in Harvey's "Nereis." But a more careful and extensive study of it, by Dr. Farlow, has convinced him that it is quite a distinct species, and he has named it for a celebrated French Algologist, Prof. Ed. Bornet.

This plant has a delicate, slender, filiform frond, consisting of a single series of naked, pink cells, placed end to end. It branches by regular forkings, and the branches are composed the same as the stem of a series of single cells. The forking is accomplished by two cells, starting from the top of one. The branches repeatedly fork in the same way, nar-

* Griffithsia. Named for Mrs. Griffith, a celebrated **English Algologist.**

rowly, till it comes about that there is quite a bushy, fan-shaped, level-topped plant, all derived from the simple beginning of a slender, single-celled thread.

It grows on *Zostera*, and other plants below tide marks. It has a beautiful rosy color, is very soft and fragile, and adheres firmly to paper. In mounting, it should not at first be put under much pressure; nor should it be "floated" in fresh water, else it will discharge its pink color. Miss Booth finds it in abundance at Orient, in July and August. It will be found on most shores south of Cape Cod. If it occurs at all in the waters of Massachusetts Bay, it must be as a great rarity, for neither my correspondents nor myself have ever found it there.

Genus.— *CALLITHAMNION,** *Lyngb.*

This is a large genus, of very beautiful plants, numbering over twenty species in our flora. In structure, they are the simplest of the red Algæ, and have what is deemed to be the most primitive method of reproduction. The frond consists of a series of single cells, put end to end, stem and branches being alike in this regard. In some species, however,

* Callithamnion = A beautiful shrub.

the main stem is more or less coated towards the base, by a covering of small cells.

It comes within the purpose of this book to direct attention only to those few species, which are specially notable for their beauty, their plentifulness, or their wide distribution. Standing at the head of the list, of our Atlantic *Callithamnia*, in respect to beauty, if not at the head of the genus itself in that regard, is

CALLITHAMNION AMERICANUM, HARV.

This plant grows not uncommon along the whole coast, from Halifax to New York. In the warmer waters, south of Cape Cod, it seems to be of a finer and more delicate habit, as well, also, as of a more brilliant rose-red color, than in the north. It is among the earliest plants to be found. I have most exquisite specimens, collected by my friend, A. R. Young, about New York, as early as March 12th. And he assures me that he has found it in fine development among the ice, on Washington's Birthday.

In Plate XX., the artist has reproduced, with great faithfulness and spirit, one of the plants of this species, with which Mr. Young has enriched my collection. It will convey some hint, I hope, of the beauty of this wonderful plant. But I believe a

.

somewhat detailed description will not be quite superfluous.

The frond is of cobweb fineness; about three or
four inches high, densely tufted, much and finely
branched; the primary branches long; the secondary
alternate and decompound, all rather widely spreading; somewhat far asunder at the base; more closely
crowded toward the top. A marked characteristic
of this and the next species is the presence, along
all the branches, primary and secondary, springing
from the top of each joint, of a *pair* of *much-
divided ramuli*, one-tenth of an inch long or more,
standing out widely from the branches.

They are easily seen with the naked eye, and
under a glass appear to be divided into long and
extremely fine branches. The joints of these fine
divisions of the ramuli are eight or ten times longer
than broad. This will serve to distinguish them from
the ramuli of the next species, the joints of which
are short and stout. It grows in deep water on
shells, stones and rocks. Mr. Collins has collected it
as late as June, at Revere, and Mrs. Davis reports
it very plentiful, in the spring, at Gloucester.

CALLITHAMNION PYLAISÆI, MONT.

In many respects, this is closely **related to** the

17

last species. Indeed, you will find plants, which, though easily distinguished from the extreme ˙ forms of either species, are very difficult to locate, and you will often find it no easy matter to determine to which species you will refer them.

But, in a general way, it may be said that this species is coarser than the last. Its main branches are thicker, and its secondary and further ramifications shorter. There are also particular distinguishing marks. The ramuli of this species spring from *just below the top of the joint;* they divide by *opposite* branching; they are much stouter and shorter than the ramuli of the other species, and the cells of these ramuli are much shorter, being not more than ˙ twice as long as wide.

The color, also, of this species is considerably darker than in *C. Americanum.* The plant grows to the height of three or four inches, is four or five times alternately decompounded, the branches remote towards the base, crowded at the top. It is a spring plant, growing in deep water, the same as *C. Americanum,* and has nearly the same geographical range, with a tendency to favor the northern localities.

Mr. Collins finds it at Revere, from March to May, not very common. Mrs. Bray reports it very common at Magnolia, during the same months. Mrs.

Davis finds it in Gloucester, as late as July. And Miss Booth, in August, at Peconic Bay, and Prof. Eaton, in Eastport, Maine, in August and September.

CALLITHAMNION FLOCCOSUM, AG.

This species is reported only in our northern waters, from Boston Bay northward. It is a very slender, remotely, much branched plant, very flaccid, and from four to six inches high. At the base, the branches are half an inch apart, but more crowded towards the top. This fact, together with the flaccid nature of the frond, makes the ramuli gather in flocculent masses at the ends of the secondary branches. This gives the plant a very uneven appearance.

The main stems of the tuft are most frequently twisted together into a little rope. The tops of the cells in the branches and branchlets just below where they join the cell above, are armed with a *single pair* of *opposite ramuli*. These are from one-twentieth to one-tenth of an inch long, simple or *unbranched*, spine-like, slender and sharp. This fact very readily distinguishes this species from either of the foregoing, whose ramuli are much branched.

Several marks distinguish it from the next species, *C. cruciatum*, viz.: its larger size; its different geographical habitat; and the fact of its having but a single pair of ramuli, at each joint, while *C. cruciatum*, frequently has two. The color of this species is like that of *C. Pylaisæi*, a bright red.

Mrs. Davis and Mrs. Bray find it in abundance at Niles Beach, Magnolia, during April and May. Profs. Verrill and Eaton found it common, growing on *Ptilota plumosa*, at Dog Island, Maine, and on mussel shells among the wharves at Eastport, during August and September.

CALLITHAMNION CRUCIATUM, AG.

This species grows only on the south side of Cape Cod, and is certainly somewhat scarce. It grows in deep water, on muddy rocks, in globose tufts, an inch or more high, of a bright red color; filaments, like most of the genus, very slender. The frond divides or forks not widely, the lower divisions are far apart, the upper close together. The branches themselves fork one or more times.

The ramuli, which are set in one or two pairs upon the upper end of each of the cells in the filaments, are mostly long and branched, one-twelfth of an inch long. They stand out almost perpendicular to the the axis of the filament.

The one point which distinguishes the mounted plant so that it can hardly fail of easy recognition, is

the fact that at the end of every branch the ramuli
crowd together and make a little dense or thickened
mass, giving the branch an appearance not unlike that
of a minute peacock's feather,— the pinnæ standing
a little apart all along the rachis, and then gathering
close about the end, form the well-known "eye" of
the miniature feather. There is certainly something
like this in a well-mounted specimen of *C. cruciatum.*
It is a summer plant. Miss Booth reports it not
common in August, at Orient. I have never collected it.

CALLITHAMNION BAILEYI, HARV.

This plant, which is certainly very common all
through the waters of southern New England and New
York, is by no means rare in Massachusetts Bay. It
is a well marked species, and cannot easily be confounded with any other *Callithamnion* of our coast.
It will usually be found two, or at most, three inches
high, and of a pyramidal outline.

It has a stout stem, larger than a bristle, which
runs quite through the plant to the top. From all
sides of this there spreads out widely, a series of stout
branches, longest at the base of the plant, but getting
rapidly shorter as we approach the top. This gives
the plant its pyramidal form. If separate branches are

now examined, it will be found that they repeat the
habit of the whole plant, sending out branchlets all
about, which are longer towards the lower part of the
branch, and shorter upwards.

This gives every main branch a sharply pointed
outline. These points thrust themselves out beyond
the principal mass of the frond in a very characteristic
way.   So marked is this feature, that it constitutes
the one easily recognized sign, when taken in connection with the robust stem and main branches, by
which to know the species.   Though the stem and
branches are so stout, for a *Callithamnion*, the ultimate
ramuli are very fine, short, and much alternately
divided.

The color is a fine dark red. Mr. Collins found
it at Revere, growing on *Zostera*, in September. I
have found it in abundance, all through the season,
on the south coast of New England, but strange
to say, during several seasons of diligent collecting,
have never found it at Marblehead. Miss Booth
collects it at Orient, L. I., washed ashore from deep
water.

There is no reason to regret that Professor Bailey's
name and memory have been preserved in so charming, and so well characterized a species, as is this
" beautiful little shrub."

CALLITHAMNION BORRERI, AG.

This, and the two following species, may not be so easily made out, and distinguished from each other at first, as those already described. Yet, when they are once known, the distinguishing points will be easily recognized. The geographical range of this species, on our coast, is limited to the waters on the south shores of New England and New York.

It grows in dense, soft tufts, two or three inches high. The frond is of capillary fineness, the branches long and widely spreading, the lower half of the branches mostly bare, the upper half divided and subdivided, alternately, many times, the ultimate branch-lets being long and slender, and not unfrequently turned back in graceful curves. The little plumes which the ultimate branchlets form, are made by arranging the ramuli on the two sides of the branch, like the pinnæ of a fern along its rachis or stalk.

The color is a fine, brilliant red. I have collected it in summer and late fall, at Newport and Wood's Holl. Miss Booth found it not very plenty at Orient in August, washed ashore from deep water.

CALLITHAMNION BYSSOIDES, ARN.

Beginners will more easily confound this species with the last, than with any other, and yet it differs from it in several well marked particulars. It is much finer in all its parts, and shows to the naked eye no main stem and branches, which are much thicker than the ultimate ramifications. To be sure, the general habit of the plant and the method of branching is much the same as that of *C. Borreri*, but the ultimate ramuli are no more than half as long, or as thick. Indeed, the whole plant is almost as fine as a spider's thread.

The color is a less brilliant red than that of *C. Borreri*, and approaches much nearer that of *C. corymbosum*, a dark or brownish red. But it will not be confounded with the latter, for that is a coarser plant even than *C. Borreri*.

The plant grows to the height of two or three inches, in dense tufts. As above indicated, it is excessively fine and flaccid, collapsing into a clot when drawn from the water. No leading stem or branches will be easily detected in the mounted plant, without the aid of a glass. But the various directions which the main branches take will be easily seen by the finely pinnated ends, which they put out beyond the principal mass of the frond, forming beautiful little plumules, or the tops of pyramids.

It grows during the summer upon *Zostera*, and other sub-marine plants and rocks, below low-tide.

It may be looked for along the coast, from New York to Massachusetts Bay, though I have collected it only at Wood's Holl and New York Bay. I have specimens from Narragansett Pier. It is not a very common plant, though Harvey says it may be found in several places in New York Harbor, from Hell Gate to Fort Hamilton.

CALLITHAMNION VERSICOLOR, AG.

This beautiful little *Callithamnion*, represented in Fig. 1, Plate XVIII., has all the delicate and cobweb fineness of filament which characterizes the last species. But it may be easily distinguished from that and every other species of *Callithamnion*, by the peculiarity which its name indicates, viz.: its striking and beautiful diversity of color. Some parts of the frond will be a brilliant rosy red, while others are an equally brilliant, full green. Sometimes a branch will begin a red and end a green, or a brown, or a yellow. Again, some one of the secondary branches on a primary will be all red, and another just by the side of it, will be a green or a yellow, and so on. Sometimes fully half a dozen different colors or shades will appear in the same frond, and I have them where the whole plant is as brilliant a green as an *Ulva* or an *Enteromorpha*.

This plant grows from one to three inches high. It has a somewhat robust leading stem with several stout primary branches, differing in this respect from *C. byssoides*, but the final branchlets and ramuli are extremely fine and delicate, and somewhat long.

A variety of this species, *seirospermum*, differs from the typical form by being a trifle stouter and coarser, with the ultimate ramuli not so abundant or so long and silky. It has, however, much the same habit of growth, and with the aid of a good lens, may be determined without difficulty, when in fruit, by the singular strings of bead-like spores which it produces in the place of the common, asexual tetra-spores. The tetraspores of this genus grow externally up the ultimate ramuli.

This species is reported from New York north-ward, but it cannot be common in northern waters, for none of my correspondents have found it in that region. But it is not very rare south of Cape Cod. I have taken numbers of plants, var. *seirospermum*, at Wood's Holl, in July. Miss Booth gathered the same at Orient, in July. I have a number of ex-quisite plants of the normal form, sent to me by Mrs. Woodward, from Cottage City, Martha's Vine-yard. I understand them to be winter plants. One of them is represented in Plate XVIII.

CALLITHAMNION CORYMBOSUM, AG.

There are very very few more beautiful plants in the sea than this. Carefully laid out, each separate plant upon a paper by itself, it may well claim to rival almost any other for gracefulness of outline, regularity and beauty of branching, and fineness and delicacy of filament.

It grows upon *Zostera*, and upon the mud-covered rocks, and piles about the docks, and along the shores, below tide, in little globose tufts, one to two and one-half inches high. Each separate plant in the tuft grows from a minute disk, with a single main stem not much thicker than a hair. This throws out widely, long branches from every side. These branches are bare at the base, but soon branch in the same manner as the main stem, with second-ary branches, which are also bare at the base, and rapidly divide and sub-divide towards the top.

The ultimate ramuli are very fine and level-topped, so as to make a great number of minute corymbs at the extremity of the branches, hence the name of the species. The general aspect of the plant is that of a miniature, bushy, very symmetri-cal shrub, the pyramidal outline of the end of the branches appearing beyond the general mass. Fig. 1, Plate XIX., gives a very excellent representation of it.

In the water, it is often a deep, rich red, but when on paper the red has a marked brown shade. It is common along the whole coast from New York northward, from June to November. I have collected it in abundance on *Zostera*, in Marblehead Harbor, in August, and on the piers at Wood's Holl, the very last days of October. Mr. Collins has found it in November, at Nahant.

CALLITHAMNION DASYOIDES, AG.

This and the following species are all that I shall undertake to describe of the *Callithannia* of California. This plant is more robust than any of the genus growing in the Atlantic waters. It attains a height of four inches or more. Its main stem is twice as thick as a bristle, regularly and alternately branched along its opposite sides.

These branches are of irregular length. Some of them as long as the main stem. Some half, and some a quarter as long. The primary branches also branch along the two sides in the same plane and in the same manner as the main stem. Likewise the secondary and tertiary branchlets sometimes, so that the plant becomes pinnately decompounded three or four times, the ultimate ramuli being very fine, and sometimes long.

It is scarce at Santa Barbara, from January to August, on the beach, growing parasitical on *Micro-ladia* and *Ceramium rubrum.* It is not uncommon at Santa Cruz all the season, parasitical on *Ptilota lensa.* It adheres well to paper, and the younger and smaller plants are certainly very beautiful, and well worth looking for. The color in them is a deep, rich red, of a darker shade in the older plants. I suppose it may be expected in greater abundance farther north. It is no doubt often collected at the Golden Gate

CALLITHAMNION HETEROMORPHUM, AG.

This is by far the most beautiful of the California *Callithamnia.* It is represented in Figure 2, Plate XVIII. It has a leading stem which extends through the whole plant, giving off alternate branches from two opposite sides at regular intervals. These branches shorten towards base and apex from the middle, where, in a plant two inches high, they are half an inch long. This gives the frond a very perfect lanceolate outline. From the primary, spring secondary branches in the same way, which divide alternately towards the top, in very short branchlets.

The peculiar mark of the species is the little circlet of delicate plumes which adorns the top of every joint, in the stem and branches, from the base to the end of the remotest divisions. Except on the main stem these plumes are scarcely discernible separately to the naked eye. But under a pocket lens they are easily seen, and it is these which give the plant its delicate, feathery appearance. This is a somewhat rare plant, though it is reported along the whole California coast, growing at all seasons, upon other Algæ, in pools or below tide. It is certainly well worth a long and laborious search, to fill one's hands with the fronds of this wonderful little beauty.

*   *   *   *   *

Thus ends our brief, and as it seems to me, altogether inadequate survey, of the " Sea Mosses," of our two far-parted shores. I may be permitted to hope, perhaps, that even the imperfect acquaintance which this little book shall give its readers, with these lower forms of Ocean Life, may teach, at least, the one lesson of patience and trust towards God, which the Poet learned from them, long years ago.

SEA WEED.

Not always unimpeded can I pray,
Nor, pitying saint, thine intercession claim;
Too closely clings the burden of the day,
And all the mint and anise that I pay
But swells my debt and deepens my self-blame.

Shall I less patience have, than Thou, who know
That Thou revisit'st all who wait for Thee,
Nor only fill'st the unsounded deeps below,
But dost refresh with punctual overflow
The rifts where unregarded mosses be?

The drooping sea weed hears, in night abyssed,
Far and more far the wave's receding shocks,
Nor doubts, for all the darkness and the mist,
That the pale shepherdess will keep her tryst,
And shore-ward lead again her foam-fleeced flocks.

For the same wave that rims the Carib shore
With momentary brede of pearl and gold,
Goes hurrying thence to gladden with its roar
Lorn weeds bound fast on rocks of Labrador
By love divine on one sweet errand rolled.

And, though Thy healing waters far withdraw,
I, too, can wait and feed on hope of Thee
And of the dear recurrence of Thy law,
Sure that the parting grace that morning saw
Abides its time to come in search of me.

J. R. Lowell

A SEA VIEW.

I climbed the sea-worn cliffs that edged the shore,
And looking downward watched the breakers curl
Around the rocks, and marked their mighty swirl
Quiver through swaying sea weed dark and hoar.
Eastward the white caps rose with far-off roar,
Against a sky like red and purple pearl,
Then hollowed greenly in, and rushed to hurl
Their weight of water at the cliffs before.
Only a sea-gull flying silently,
And one soft rosy sail were now in sight,—
A sail the sunset touched right tenderly,
And flushed with dreamy glory faintly bright.
Then fain would I have crossed the tossing sea.
Fain dared the storm to float within that light.

Alice Osborne.

GLOSSARY.

—:o:—

ALGA, *Plural* ALGÆ.	Cryptogamic plants which grow in the water.
ARTICULATED.	Jointed.
AXIL.	The angle. on the upper side, between the branch and the stem, or between two branches.
AXIS.	The central line, or direction, of the main body of the plant.
CAPILLARY.	Hair-like, in size and shape.
CARTILAGINOUS.	Firm and tough, in texture.
CILIA.	Short, slender processes, like eye-lashes.
CHLOROPHYL.	The green cell contents.
CLUB-SHAPED.	Tapering below, blunt above.
COMPRESSED.	Flattened on opposite sides; parts commonly quite narrow in Algæ.
CONCEPTACLE.	The vessel which contains the true fruit, in the Red Algæ.
CORIACEOUS.	Leathery, tough.
CORYMB.	A sort of flat or convex flower cluster; imitated in some Algæ by the ultimate ramuli at the ends of the branchlets.

CRYPTOGAM. A flowerless plant.

CYLINDRICAL. { Formed like stems generally, round, and tapering if at all, very slightly.

FILIFORM. { Thread-shaped, long, slender and cylindrical.

FLORA. { The plants of a district, or country, taken together.

FROND. { The whole body of the Alga, main stem, branches and ramuli, all taken together.

GELATINOUS. Jelly-like.

HABITAT. The place of growth of a plant.

HOLD-FAST. { The part of an Alga, which answers to the root of other plants, that by which it is attached to whatever it grows upon; it may be a mass of root-fibres, or a thin, disk-like expansion of the substance of the frond.

LANCEOLATE *or* LANCE-SHAPED. { Leaflets several times longer than wide, tapering upwards, or both upwards and downwards.

LATERAL. From the side.

LOBE. A segment of a membraneous frond.

MEMBRANEOUS. { Thin, more or less translucent, like a membrane.

MIDRIB.	A large vein, or continuation of the stalk, running through the middle of some flattened or membraneous fronds
PALMATE.	Shaped like the hand, with the fingers extended.
PETIOLE.	A leaf-stalk.
PAPILLA, *Plural* PAPILLÆ.	Little nipple-shaped protuberances.
PINNA, *Plural* PINNÆ.	Primary leaflets or branchlets of a pinnate frond.
PINNULE, *Plural* PINNULÆ.	Secondary, or still smaller leaflets or branchlets of a pinnate frond, growing on the pinnæ.
PINNATE.	Where the secondary parts are arranged along the sides of their primaries, in same regular order, opposite or alternate, like leaflets along the sides of a common petiole.
RACHIS.	That portion of the stem, along which the branches are arranged like ribs along a backbone.
RAMULUS, *Plural* RAMULI.	The smaller branches, or branchlets.

SEGMENTS.	Divisions of the fronds.
SERRATED.	Toothed like a saw.
SINUOUS.	The margin crooked, bending in and out.
SPINDLE-SHAPED.	Tapering to each end from a thickened middle.
SPINES.	Small, thorn-like processes.
SPORES.	The seeds of the Algæ, and other Cryptogamic plants.
TETRASPORES, or TETRAGONIDIA.	The asexual spores of the Red Algæ, usually arranged in groups of fours.
TOP-SHAPED.	Like a top, or a cone with the apex downwards.
TUBERCLE.	A small excrescence.
VEINS.	Small, linear thickenings of the frond, which resemble the veinings, or framework of the leaves of trees.
VESICLE.	A bladder.
WHORL.	Ramuli arranged in a circle around the stem or branches.

INDEX OF GENERA AND SPECIES.

—:o:—

278   *INDEX.*

www.ingramcontent.com/pod-product-compliance
Lightning Source LLC
Chambersburg PA
CBHW021402210326
41599CB00011B/981